LIGHTING & MAINTENANCE

PROFESSIONAL REFERENCE

Paul Rosenberg

Created exclusively
for DEWALT by:

PAL publications®

www.palpublications.com
1-800-246-2175

Titles Available From DEWALT

DEWALT Trade Reference Series

Construction Professional Reference

Datacom Professional Reference

Electric Motor Professional Reference

Electrical Estimating Professional Reference

Electrical Professional Reference

HVAC Professional Reference

Lighting & Maintenance Professional Reference

Plumbing Professional Reference

Referencia profesional sobre la industria eléctrica

Security, Sound & Video Professional Reference

Wiring Diagrams Professional Reference

DEWALT Exam and Certification Series

Electrical Licensing Exam Guide

HVAC Technician Certification Exam Guide

This Book Belongs To:

Name:____________________________

Company: ____________________________

Title: ____________________________

Department: ____________________________

Company Address: ____________________________

____

____

Company Phone: ____________________________

Home Phone: ____________________________

Pal Publications, Inc.
374 Circle of Progress
Pottstown, PA 19464-3810

First edition published 2005

NOTICE OF LIABILITY

Information contained in this work has been obtained from sources believed to be reliable. However, neither Pal Publications nor its authors guarantee the accuracy or completeness of any information published herein, and neither Pal Publications nor its authors shall be responsible for any errors, omissions, damages, liabilities or personal injuries arising out of use of this information. This work is published with the understanding that Pal Publications and its authors are supplying information but are not attempting to render engineering or other professional services. If such services are required, the assistance of an appropriate professional should be sought. The reader is expressly warned to consider and adopt all safety precautions and to avoid all potential hazards. The publisher makes no representation or warranties of any kind, nor are any such representations implied with respect to the material set forth here. The publisher shall not be liable for any special, consequential, or exemplary damages resulting, in whole or part, from the readers' use of, or reliance upon, this material.

ISBN 0-9759709-5-X

09 08 07 06 05 5 4 3 2 1

Printed in the United States of America

A Note To Our Customers

We have manufactured this book to the highest quality standards possible. The cover is made of a flexible, durable and water-resistant material able to withstand the toughest on-the-job conditions. We also utilize the Otabind process which allows this book to lay flatter than traditional paperback books that tend to snap shut while in use.

Lighting & Maintenance Professional Reference is not a substitute for the National Electric Code®. National Electric Code® and NEC® are registered trademarks of the National Fire Protection Association, Inc. Quincy, MA.

Preface

Electrical Maintenance is a broad and complicated field. A maintenance technician must be competent in a wide variety of disciplines. In addition, he or she must also be adept at scheduling and must possess a variety of managerial skills.

I have divided this book into two sections. The first section deals with lighting and the second references important electrical and general maintenance information.

If you are familiar with the DEWALT Trade Reference Series, you will notice that some of the chapters in this book contain more explanatory information than usual. I felt it was important to include it here so that you are properly informed on some of the more unusual and technical subjects. This is especially true for the section on lamp disposal, which has been regulated by the government during the past few years. Violations of these regulations can lead to stiff penalties.

This book contains material that is necessary for anyone in the maintenance field to have in their possession at all times. Naturally, a topic may have been overlooked or not discussed at a depth suitable for all maintenance workers. I will constantly monitor and update this book on a regular basis to include additional material requested by our readers and to modify the book according to the development of the maintenance industry.

Best wishes,
Paul Rosenberg

CONTENTS

CHAPTER 2 – *HIDs*. 2-1

CHAPTER 4 – *Incandescents* 4-1

CHAPTER 5 – *Lamp Disposal* 5-1

CHAPTER 9 – *Motor Maintenance* . . . 9-1

CHAPTER 1
LIGHTING & LAMPS

Lighting is the most widely-used electrical technology and the second greatest energy user in any building. It is serviced by nearly all maintenance workers. The following are explanations of the most basic lighting technologies:

LIGHT OUTPUT TERMINOLOGY

The most common measure of light output is the *lumen (luminous flux)*. Light sources (lamps) come labeled with an output rating measured in *lumens*. For example, a T12 40-watt fluorescent tube typically has a rating of 3050 lumens. Also, the output of an entire light fixture can be expressed in lumens. As lamps and fixtures age, their lumen output decreases; that is, the amount of light they put out goes down. This is called *lumen depreciation*. Most lamp ratings are based on *initial lumens* – the output of new lamps.

LIGHT LEVELS

The intensity of light is called *illuminance*, and is measured in *foot-candles* (lumens per square foot). Foot-candles are measured with standard light meters. Hold them on the work surface where tasks are performed, and keep the meter away from your body or other reflective surfaces. For planning purposes, use a manufacturer's photometric data to reliably predict foot-candle levels in a defined area.

ELECTRICITY-TO-LIGHT EFFICIENCY

Different types of light production are more or less efficient in converting energy into visible light than others. This is referred to as the *efficacy* of the light source, which compares the number of lumens leaving the lamp versus the number of watts required.

Efficacy is measured in *lumens per watt*. For example, H10 lamps put out far more lumens per watt than incandescent lamps, and thus have a higher efficacy.

LIGHT COLOR TEMPERATURE

This is a measurement of "warmth" or "coolness" provided by the lamp. Color temperature is expressed in degrees Kelvin. For example, an electric stove element changes in color as its temperature increases; first red, then orange, yellow, and finally bluish white. A "warm" light source refers to a more yellow or orange tint of light, even though it has a lower color temperature. Likewise, a "cooler" light source has more of a bluish tint, even though the color temperature is higher. For example, a cool-white fluorescent lamp appears bluish in color with a color temperature of around 4100 K. A warmer fluorescent lamp appears more yellowish with a color temperature around 3000 K.

COLOR RENDERING INDEX

Color rendering index (CRI) is a relative scale, ranging from 0-100, indicating how well perceived colors match actual colors. It measures the degree that perceived colors of objects, illuminated by a given light source, conform to the colors of those same objects when they are lighted by a reference standard light source. The higher the color rendering index, the less color shift or distortion occurs.

The CRI number does not indicate which colors will shift or by how much; it is rather an indication of the average shift of eight standard colors. Two different light sources may have identical CRI values, but colors may appear quite different under these two sources. So, the CRI is a useful guideline, but is not a perfect scientific indicator.

CRIs in the range of 75-100 are considered excellent, while 65-75 are good. The range of 55-65 is fair, and 0-55 is poor. Under higher CRI sources, surface colors appear brighter, creating the illusion of higher illuminance levels.

DETERMINING TARGET LIGHT LEVELS

The Illuminating Engineering Society of North America (often called the IES) publishes recommended light levels which include the following factors:

- The task(s) being performed (contrast, size, etc.).

- The ages of the occupants.
- The importance of speed and accuracy.

Once the lighting level has been determined, you can select the best combination of lamps and fixtures based on the following criteria:

- Fixture efficiency.
- Lamp lumen output.
- The reflectance of surrounding surfaces.
- The effects of light losses from lamp lumen depreciation and dirt accumulation.
- Room size, shape and availability of natural light.

When you design a lighting system (new or upgraded), you must be careful to avoid overlighting a space. This wastes energy and may not provide any better results.

Years ago, some offices were designed for 100 foot-candles or more, where 50 foot-candles were adequate or even superior. For example, to light a space that uses computers, the overhead fixtures should provide up to 30 fc (foot-candles) of ambient lighting. Task lights should provide the additional foot-candles needed to achieve a total illuminance of up to 50 fc for reading and writing. (Task lights are small, direction light fixtures such as reading lamps and under-cabinet fluorescent strips.)

LIGHTING QUALITY

In addition to the color rendering index, there are other important ways to measure lighting quality.

Visual comfort probability (VCP) indicates the percent of people who are comfortable with the glare from a fixture.

Spacing criteria (SC) refers to the maximum recommended distance between fixtures to ensure uniformity of light and to avoid glare.

LIGHT SOURCES

Selecting the appropriate source depends on installation requirements, life-cycle cost, color qualities, dimming capability and the effect wanted.

Electric light sources have three characteristics: efficacy, color temperature and color rendering index.

QUALITY LIGHTING MEASURES

- **Visual comfort probability (VCP)** indicates the percent of people who are comfortable with the glare from a fixture.
- **Spacing criteria (SC)** refers to the maximum recommended distance between fixtures to ensure uniformity.
- **Color rendering index (CRI)** indicates the color appearance of an object under a source as compared to a reference source.

QUANTITY LIGHTING MEASURES

- **Luminous flux** is commonly called light output and is measured in lumens (lm).
- **Illuminance** is called light level and is measured in foot-candles (fc).
- **Luminance** is referred to as brightness and is measured in footlamberts (fl) or candelas/m^2 (cd/m^2).

PRIMARY LIGHTING TERMS AND UNITS OF MEASUREMENT

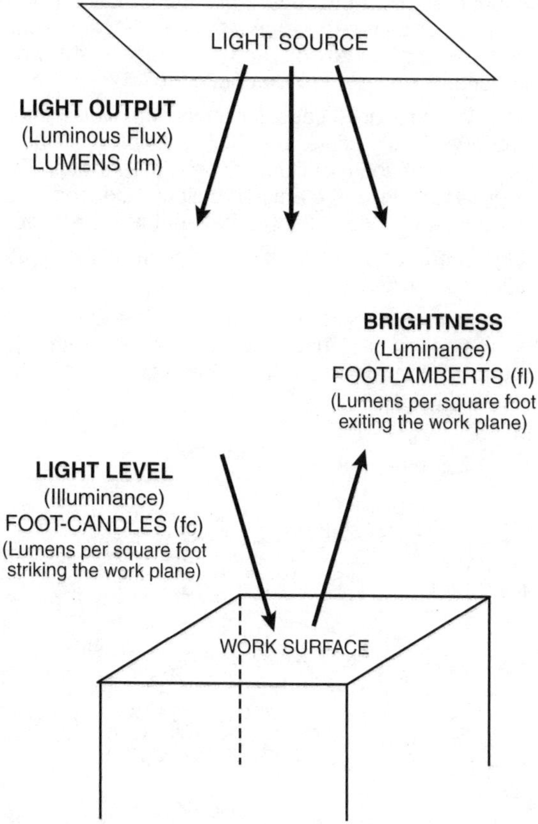

INVERSE SQUARE LAW

A *lamp* converts electrical energy into light. The amount of light produced is expressed in lumens. For example, a standard 40 watt incandescent lamp produces about 480 lm, and a standard 40 watt fluorescent lamp produces about 3100 lm.

The light produced causes illumination. *Illumination* is the effect that occurs when light falls on a surface. The unit of measure of illumination is the foot-candle. A *foot-candle* (fc) is the amount of light produced by a lamp (lumens) divided by the area that is illuminated.

Light spreads as it travels farther from the light-producing source.

The *inverse square law* states that the amount of illumination on a surface varies inversely with the square of the distance from the light source as illustrated below.

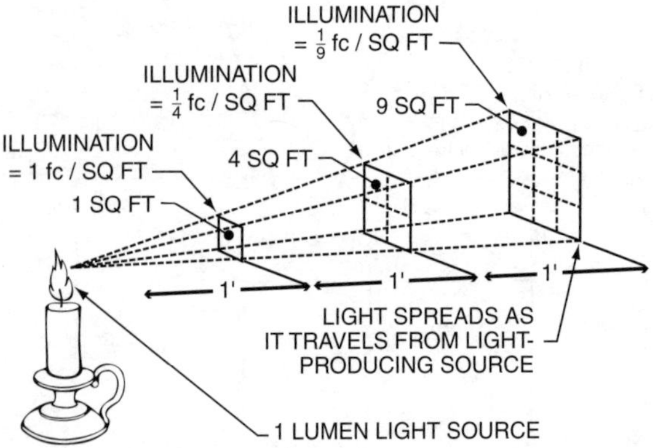

LUMEN AND FOOT-CANDLE FORMULAS

Lumens required per individual lamp

$$\text{Lumens} = \frac{\text{foot-candles desired} \times \text{area per lamp (sq. ft.)}}{\text{coefficient of utilization} \times \text{depreciation factor}}$$

Foot-candles required for computing lamps of various sizes

$$\text{Foot-candles} = \frac{\text{lamp lumens} \times \text{coefficient of utilization} \times \text{depreciation factor}}{\text{area per lamp (sq. ft.)}}$$

CALCULATING MAINTAINED LIGHT LEVEL

$$\textbf{fc} = \frac{\text{rated lumens} \times \text{CU} \times \text{BF} \times \text{LSD} \times \text{RSDD} \times \text{LLD} \times \text{LDD} \times \text{LBO}}{\text{area of room (ft}^2\text{)}}$$

CU	=	coefficient of utilization
BF	=	ballast factor
LSD	=	luminaire surface depreciation
RSDD	=	room surface dirt depreciation
LLD	=	lamp luminaire depreciation
LDD	=	luminaire dirt depreciation
LBO	=	lamp burnouts (%)

COMMON UNITS OF BRIGHTNESS

1 Candle per square inch	=	452 footlamberts
	=	0.487 lamberts
	=	487 millilamberts
1 Footlambert	=	1 lumen per sq. foot reflected or emitted
	=	0.00221 candles per sq. in.
	=	1.076 millilamberts
1 Lambert	=	1 lumen per sq. centimeter reflected or emitted
	=	1000 millilamberts
	=	929 footlamberts
	=	2.054 candles per sq. in.
1 Millilambert	=	0.929 footlamberts
	=	0.002054 candles per sq. in.

CONVERSION FACTORS FOR UNITS OF ILLUMINATION

Given	Multiply By	To Obtain
Illuminance (E) in lux	0.0929	foot-candles
Illuminance (E) in foot-candles	10.764	lux
Luminance (L) in cd/sq. m	0.2919	footlamberts
Luminance (L) in footlamberts	3.4263	cd/sq. m
Intensity (I) candelas	1.0	candlepower

RECOMMENDED ILLUMINATION RANGES

	Ranges of illuminances (fc)			
Occupant's Age	<40	40-55	55+	Type of Activity P.T. = Part Time F.T. = Full Time
Category	Low	Med	High	
A	2	3	5	Dark public areas
B	5	7.5	10	Simple orientation (P.T.)
C	10	15	20	Visual tasks (P.T.)
D	20	30	50	Visual tasks of high contrast (P.T.)
E	50	75	100	Visual tasks of medium contrast (F.T.)

RECOMMENDED ILLUMINATION RANGES *(cont'd)*

	Ranges of illuminances (fc)			
Occupant's Age	<40	40-55	55+	Type of Activity P.T. = Part Time F.T. = Full Time
Category	Low	Med	High	
F	100	150	200	Visual tasks of low contrast (F.T.)
G	200	300	500	Visual tasks of small size (F.T.)
H	500	750	1000	Exacting visual tasks (F.T.)
I	1000	1500	2000	Special visual tasks of extremely small size (F.T.)

RECOMMENDED LIGHT LEVELS

Interior		Exterior	
Area	**fc**	**Area**	**fc**
Assembly		**Airports**	
Rough	30	Terminal	10
Medium	100	Loading	2
Banks		**Buildings**	
Lobby	50	Light surface	15
Tellers	150	Dark surface	50
Hospital/Med.		**Construction**	
Dental	1000	General	10
Operating	2500	Excavation	2
Machine Shop		**Parking Areas**	
Rough	50	Industrial	2
Medium	100	Shopping	5
Offices		**Loading Areas**	
Regular	100	Pier	20
Detailed work	200	Trucking	30
Printing		**Service Station**	
Proofreading	150	Pumps	25
Color inspecting	200	Service	5

REFLECTANCE VALUES OF VARIOUS MATERIALS AND COLORS

Type of Material	Approximate Reflectance (in %)
Acoustical ceiling tile	75-85
Aluminum, brushed	55-58
Aluminum, polished	60-70
Clear glass	8-10
Granite	20-25
Marble	30-70
Stainless steel	55-65
Wood	
Light oak	25-35
Dark oak	10-15
Mahogany	6-12
Walnut	5-10

Color	Approximate Reflectance (in %)
White	80-85
Light gray	45-70
Dark gray	20-25
Ivory white	70-80
Ivory	60-70
Pearl gray	70-75
Buff	40-70
Tan	30-50
Brown	20-40
Green	25-50
Azure blue	50-60
Sky blue	35-40
Pink	50-70
Cardinal red	20-25
Red	20-40

LIGHT SOURCE CHARACTERISTICS

			High-Intensity Discharge			
	Incandescent, Including Tungsten	Fluorescent	Mercury-Vapor (Self-Ballasted)	Metal-Halide	High-Pressure Sodium (Improved Color)	Low-Pressure Sodium
Wattages (lamp only)	15-1500	15-219	40-1000	175-1000	70-1000	35-180
Life[a] (hr)	750-12,000	7500-24,000	16,000-15,000	1500-15,000	24,000 (10,000)	18,000
Efficacy[a] (lumens/W) lamp only	15-25	55-100	50-60 (20-25)	80-100	75-140 (67-112)	Up to 180
Lumen maintenance	Fair to excellent	Fair to excellent	Very good (good)	Good	Excellent	Excellent
Color rendition	Excellent	Good to excellent	Poor to excellent	Very good	Fair	Good
Light direction control	Very good to excellent	Fair	Very good	Very good	Very good	Fair
Source size	Compact	Extended	Compact	Compact	Compact	Extended
Relight time	Immediate	Immediate	3-10 min	10-20 min	Less than 1 min	Immediate
Comparative fixture cost	Low: simple fixtures	Moderate	Higher than incandescent and fluorescent	Generally higher than mercury	High	High
Comparative operating cost	High: short life and low efficiency	Lower than incandescent	Lower than incandescent	Lower than mercury	Lowest of HID types	Low
Auxiliary equipment needed	Not needed	Needed: medium cost	Needed: high cost	Needed: high cost	Needed: high cost	Needed: high cost

[a]Life and efficacy ratings subject to revision. Check manufacturers' data for latest information.

LIGHT SOURCE EFFICACIES

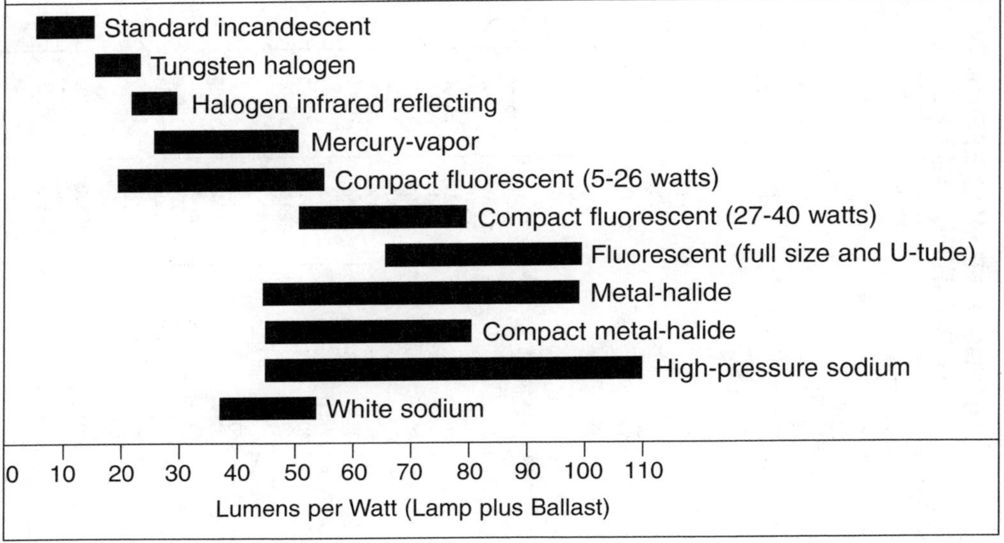

LUMENS, LIFE, AND EFFICACY FOR VARIOUS LAMPS

Lamp Type (Watts)	Initial Lumens	Life (hours)	Efficacy (lumens/watts)
Incandescent, standard inside frosted			
25	235	1000	9
40	480	1500	12
60	840	1000	14
75	1210	850	16
100	1670	750	17
150	2850	750	19
200	3900	750	19
300	6300	1000	21
500	10750	1000	21
1000	23100	1000	23
Incandescent PAR-38			
150 Spot	1100	2000	7
150 Flood	1350	2000	9
Quartz incandescent			
500	10550	2000	21
1000	21400	2000	21
1500	35800	2000	24
40	3150	20000	78
40	2200	20000	55
60	4300	12000	41
60	3050	12000	50
85	2850	20000	81
85	2000	20000	57
Fluorescent, U-line			
40	2850	12000	71
40	2020	12000	50
Fluorescent, instant start			
60	5600	12000	93
60	4000	12000	66
75	6300	12000	84
75	4500	12000	60
95	8500	12000	89
95	6100	12000	64
110	9200	12000	83
110	6550	12000	59
180	12300	10000	68
215	14500	10000	67
215	13600	10000	63

LUMENS, LIFE, AND EFFICACY FOR VARIOUS LAMPS *(cont'd)*

Lamp Type (Watts)	Initial Lumens	Life (hours)	Efficacy (lumens/watts)
Mercury-vapor			
40	1140	24000	28
50	1575	24000	31
75	2800	24000	37
100	4300	24000	43
175	8500	24000	48
175	7900	24000	45
250	13000	24000	52
250	12100	24000	48
400	23000	24000	57
400	21000	24000	52
700	43000	24000	61
1000	63000	24000	63
Metal-halide			
175	14000	7500	80
250	20500	7500	82
400	34000	15000	85
1000	105000	10000	105
High-pressure sodium			
50	3300	20000	66
70	5800	20000	82
100	9500	20000	95
150	16000	24000	106
150*	12000	12000	80
200	22000	24000	110
215*	19000	12000	88
250	27500	24000	110
400	50000	24000	125
1000	140000	24000	140
Low-pressure sodium			
65	4800	18000	137
65	8000	18000	145
90	13500	18000	150
135	22500	18000	166
180	33000	18000	183

CRI VALUES FOR SELECTED LIGHT SOURCES

Light Source	Typical CRI Value
Incandescent/Halogen	100
Fluorescent	
Cool White T12	62
Warm White T12	53
High Lumen T12	75-85
T8	75-85
T10	80-85
Compact	80-85
Mercury-Vapor (clear/coated)	22/52
Metal-Halide (clear/coated)	65/85
High-Pressure Sodium	
Standard	25
Deluxe	70
White HPS	80
Low-Pressure Sodium	0-18

BRIGHTNESS

Incorrect Lamp Spacing	Correct Lamp Spacing
Uneven Brightness	Even Brightness

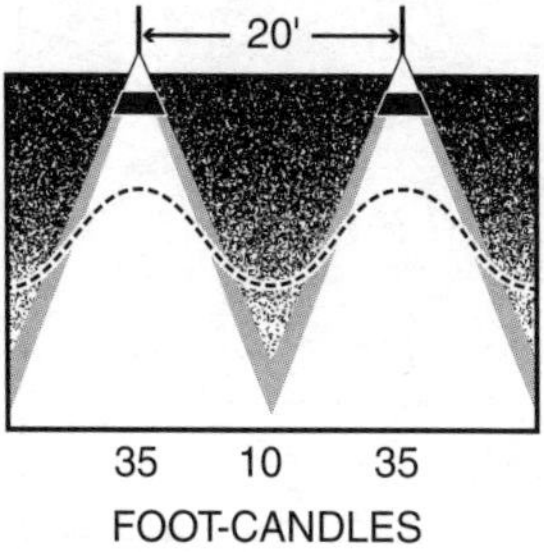

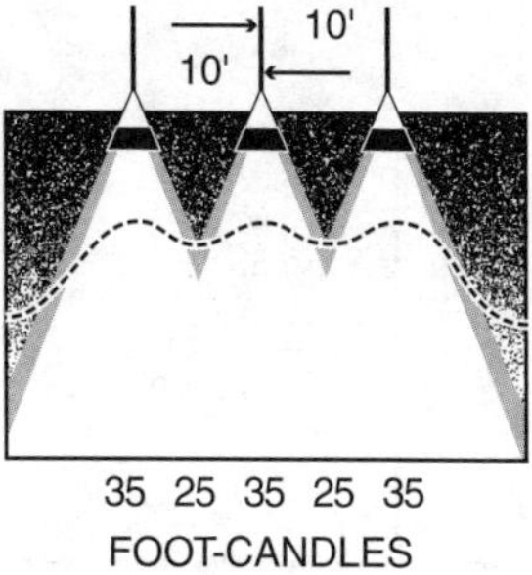

CONTRAST

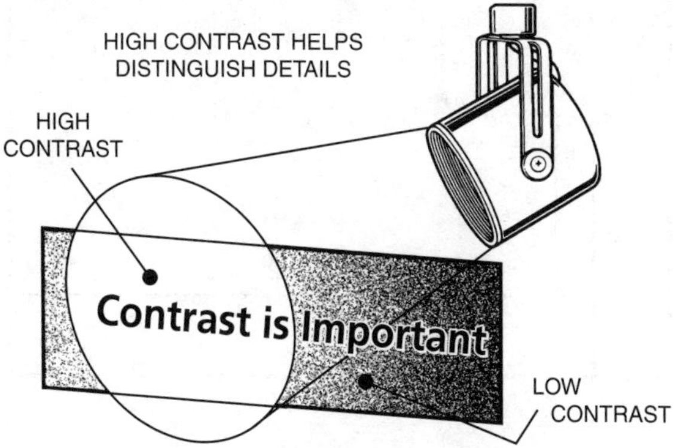

LIGHT FIXTURE SPACING (Foot-candles)

IMPROPER

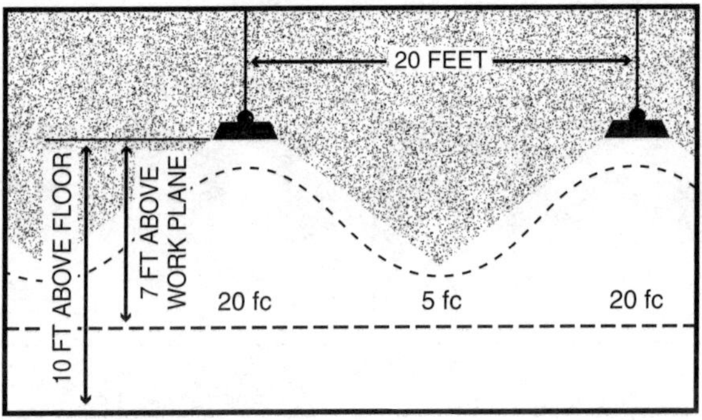

PROPER

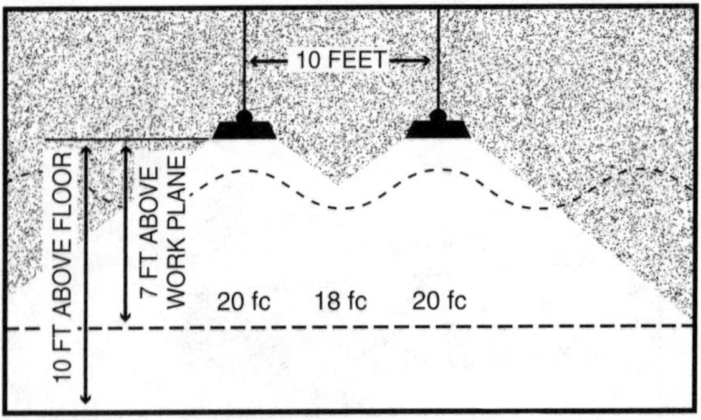

TYPICAL LIGHTING DISTRIBUTION CURVE

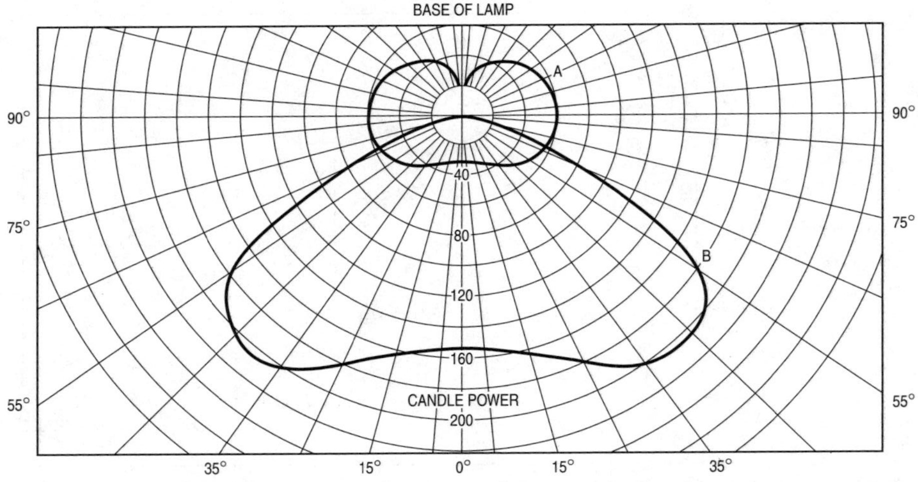

(A) Candlepower at 10 feet for tungsten lamp. (B) Candlepower at 10 feet for tungsten lamp with a reflector.

VERTICAL, PERPENDICULAR AND HORIZONTAL PLANES

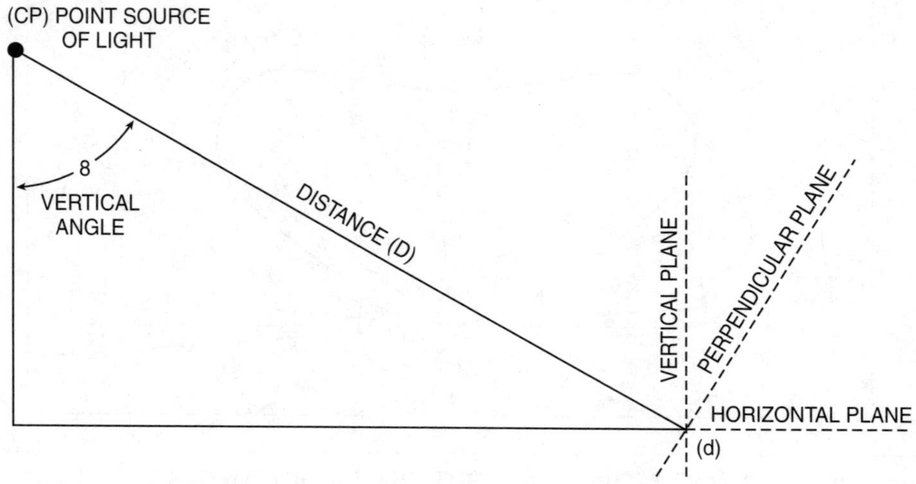

LAMP ADVANTAGES AND DISADVANTAGES

Lamp	Advantages	Disadvantages
Incandescent, tungsten-halogen	Low initial cost Simple construction No ballast required Available in many shapes and sizes Requires no warm-up or restart time Inexpensively dimmed Simple maintenance	Low electrical efficiency High operating temperature Short life Bright light source in small space Does not allow large distribution of light
Fluorescent	Available in many shapes and sizes Moderate cost Good electrical efficiency Long life Low shadowing Low operating temperature Short turn-ON delay	Not suited for high-level light in small, highly-concentrated applications Requires ballast Higher initial cost than incandescent lamps Light output and color affected by ambient temperature Expensive to dim
Low-pressure sodium, mercury-vapor, metal-halide, high-pressure	Good electrical efficiency Long life High light output Slightly affected by ambient temperature	May cause color distortion Long start and restart time High initial cost High replacement cost Requires ballast Expensive or not possible to dim Problem starting in cold weather High-socket voltage required

LAMP RATINGS (W=WATTS)

Lamp	Initial Lumen	Mean Lumen
40 W standard incandescent	480	N/A
100 W standard incandescent	1750	N/A
40 W standard fluorescent	3400	3100
100 W tungsten-halogen	1800	1675
100 W mercury-vapor	4000	3000
250 W mercury-vapor	12,000	9800
100 W high-pressure sodium	9500	8500
250 W high-pressure sodium	30,000	27,000
250 W metal-halide	20,000	17,000

LAMP CHARACTERISTICS

Characteristic	Standard Incandescent	Tungsten-Halogen
Wattage	3-1,500	10-1,500
System Efficacy *(Lumens per watt)*	6-241	8-33
Average Rated Life *(Hours)*	750-2,000	2,000-4,000
CRI	95+	95+
Life Cycle Cost	high	high
Fixture Size	compact	compact
Start to Full Brightness	immediate	immediate
Restrike Time	immediate	immediate
Lumen Maintenance	good/excellent	excellent

LAMP CHARACTERISTICS *(cont'd)*

Characteristic	Fluorescent	Compact Fluorescent
Wattage	4-215	5-40
System Efficacy *(Lumens per watt)*	50-100	50-80
Average Rated Life *(Hours)*	7,500-24,000	10,000-20,000
CRI	49-92	82-86
Life Cycle Cost	low	moderate
Fixture Size	extended	compact
Start to Full Brightness	0-5 seconds	0-1 minute
Restrike Time	immediate	immediate
Lumen Maintenance	good	good

LAMP CHARACTERISTICS *(cont'd)*

Characteristic	Mercury-Vapor	Metal-Halide
Wattage	40-1,250	32-2,000
System Efficacy *(Lumens per watt)*	25-50	50-115
Average Rated Life *(Hours)*	24,000+	6,000-20,000
CRI	22-52	65-92
Life Cycle Cost	moderate	moderate
Fixture Size	compact	compact
Start to Full Brightness	2-5 minutes	2-5 minutes
Restrike Time	3-10 minutes	10-20 minutes
Lumen Maintenance	poor/fair	good

LAMP CHARACTERISTICS *(cont'd)*

Characteristic	High-Pressure Sodium	Low-Pressure Sodium
Wattage	35-1,000	18-180
System Efficacy *(Lumens per watt)*	40-140	120-180
Average Rated Life *(Hours)*	16,000-24,000	12,000-18,000
CRI	21-80	0-18
Life Cycle Cost	low	low
Fixture Size	compact	extended
Start to Full Brightness	4-6 minutes	10-15 minutes
Restrike Time	1 minute	immediate
Lumen Maintenance	good/excellent	excellent

LAMP LUMEN DEPRECIATION

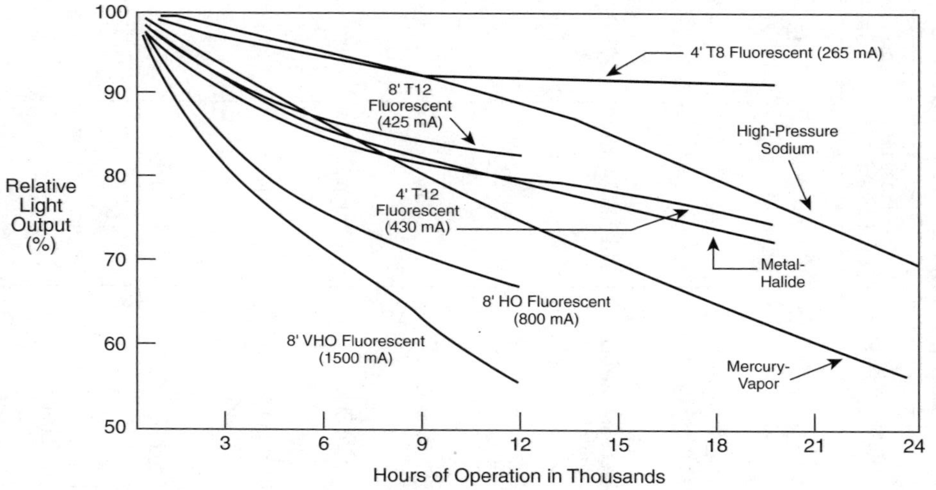

LAMP ENERGY DISTRIBUTION CURVES

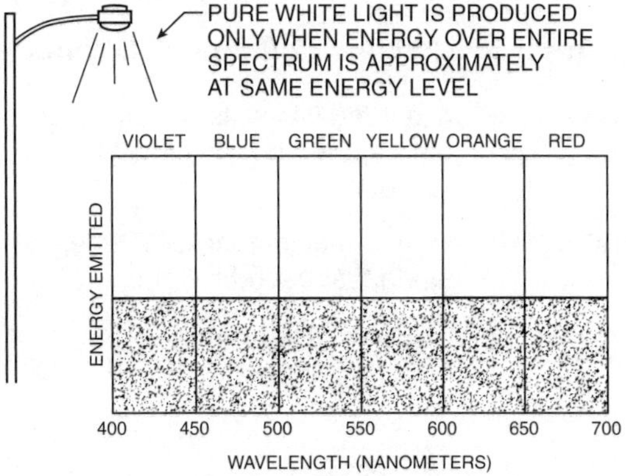

PURE WHITE LIGHT

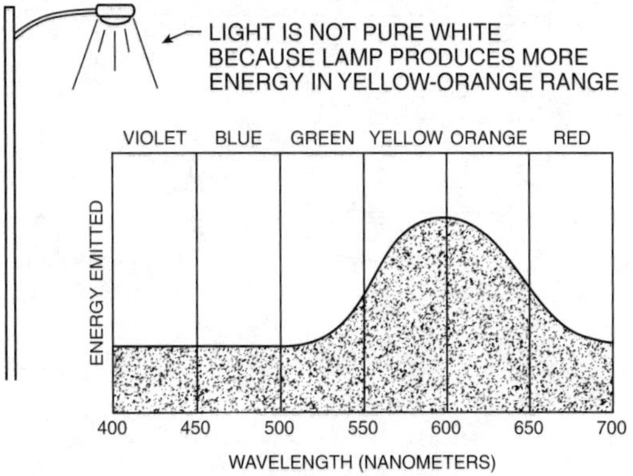

LOW-PRESSURE SODIUM LAMP

MEASURING SAVINGS FROM DIMMING CONTROLS

$$\text{kWh saved} = [(\text{kW})_{\text{before}} \times (\text{hrs})_{\text{after}}] - (\text{kWh})_{\text{after}}$$

$(\text{kW})_{\text{before}}$ = measured kilowatts of static load (before upgrade) on circuit to be metered

$(\text{hrs})_{\text{after}}$ = hours of equipment "ON" time during test period (after upgrade)

$(\text{kWh})_{\text{after}}$ = measured kilowatt-hours (after upgrade) on metered circuit

DIMMING CIRCUIT FOR SERIES-CONNECTED COLD-CATHODE LAMPS

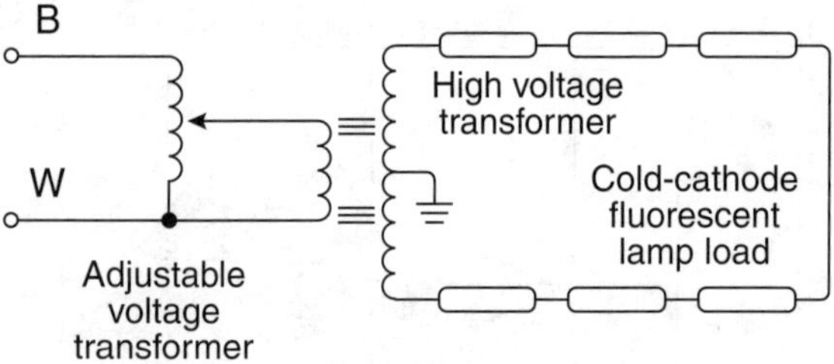

ENERGY SAVINGS POTENTIAL WITH OCCUPANCY SENSORS

Application	Energy Savings*
Offices (private)	25-50%
Offices (open spaces)	20-25%
Rest Rooms	30-75%
Corridors	30-40%
Storage Areas	45-65%
Meeting Rooms	45-65%
Conference Rooms	45-65%
Warehouses	50-75%

*Note: Figures listed represent maximum energy savings potential under optimum circumstances. Figures are based on manufacturer estimates. Actual savings may vary.

TYPICAL LIGHTING CONTROL SYSTEM

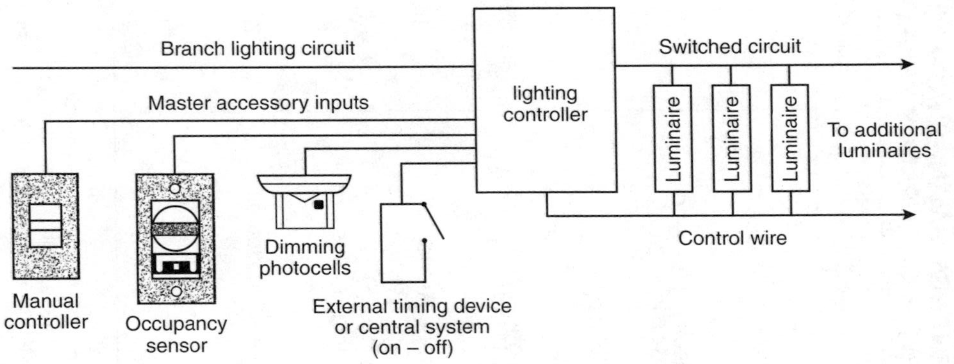

CHAPTER 2
HIDs

High-intensity discharge (HID) lamps are similar to fluorescents in that they produce light with an arc that is maintained between two electrodes. The arc in HIDs is shorter, yet generates much more light, heat and pressure within the arc tube.

Although originally developed for outdoor and industrial applications, HIDs are now used in office, retail and other indoor applications. Their color rendering characteristics have been improved and lower wattages have recently become available (as low as 18 watts).

There are several advantages to HID sources:

- Relatively long life (5,000 to 24,000+ hrs)
- Relatively high lumen output per watt
- Relatively small in physical size

However, HIDs do have limitations: First of all, they require time to warm up. This varies from lamp to lamp, but the average is 2 to 6 minutes. Second, HIDs have a *restrike* time. If there is a momentary interruption of current or a substantial voltage drop, the arc will be extinguished. At that point, the gases inside the lamp are too hot to ionize again, and time is needed for the gases to cool and pressure to drop for the arc to restrike. Restriking takes between 5 and 15 minutes, depending on the type being used. Therefore, the best applications for HIDs are areas where lamps are not switched on and off intermittently.

HID sources listed from highest to lowest efficacy:

- Low-pressure sodium (poor color rendering)
- High-pressure sodium (fair-to-good color rendering)
- Metal-halide (good-to-excellent color rendering)
- Mercury-vapor (good color rendering)

BASIC CHARACTERISTICS OF HID LAMPS

High-Pressure Sodium

- ★ This lamp uses sodium vapor under high pressure producing an orangish white light
- ★ 3-6 minutes start time
- ★ 30-60 seconds restart time
- ★ 65-115 lumens per watt
- ★ 7,500-14,000 hours of lamp life

Low-Pressure Sodium

- ★ This lamp uses sodium vapor under low pressure producing a yellow/orange light
- ★ 6-12 minutes start time
- ★ 4-12 seconds restart time
- ★ 190-200 lumens per watt
- ★ 1800 hours of lamp life

Metal-Halide

- ★ This lamp uses mercury vapor with metal halides producing a white light
- ★ 2-5 minutes start time
- ★ 10-15 minutes restart time
- ★ 80-125 lumens per watt
- ★ 3,000-20,000 hours of lamp life

Mercury-Vapor

- ★ This lamp uses mercury vapor producing a bluish white light
- ★ 5-6 minutes start time
- ★ 3-5 minutes restart time
- ★ 50-60 lumens per watt
- ★ 16,000-24,000 hours of lamp life

SPECIFIC HID LAMP INFORMATION

HID Source	Type	Lumens	Watts	Life (Hours)	Base
High-pressure sodium	B-17 clear	2,250	35	24,000	Medium
	E-23½ clear	4,000	50	24,000	Mogul
	E-23½ clear	6,400	70	24,000	Mogul
	B-17 clear	9,500	100	24,000	Medium
	B-17 clear	16,000	150	24,000	Medium
	E-23½ clear	16,000	150	24,000	Mogul
	E-18 clear	30,000	250	24,000	Mogul
	E-25 clear	140,000	1000	24,000	Mogul
Low-pressure sodium	T-16 clear	1,570	18	16,000	Bayonet
	T-16 clear	4,000	35	16,000	Bayonet
	T-16 clear	6,655	55	16,000	Bayonet
	T-21 clear	11,095	90	16,000	Bayonet
	T-21 clear	19,140	135	16,000	Bayonet

SPECIFIC HID LAMP INFORMATION *(cont.)*

HID Source	Type	Base	Watts	Lumens	Life (Hours)
Metal-halide	ED-17 coated	Medium	70	4,800	10,000
	ED-17 coated	Medium	150	12,200	10,000
	E-28/BT-28 coated	Mogul	175	14,000	10,000
	E-28/BT-28 coated	Mogul	250	20,500	10,000
	E-37/BT-37 coated	Mogul	400	34,000	20,000
	BT-56 coated	Mogul	1000	110,000	12,000
Mercury-vapor	E-23½ white	Mogul	100	4,200	24,000
	E-28 white	Mogul	175	8,600	24,000
	E-28 white	Mogul	250	12,100	24,000
	E-37 or BT-37 white	Mogul	400	22,500	24,000
	BT-56 white	Mogul	1000	63,000	24,000

COMMON HID LAMP BASES WITH ANSI DESIGNATIONS

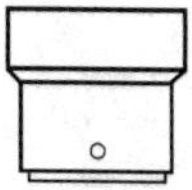

2-Lug sleeve
B22d

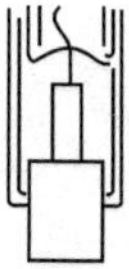

Recessed
single contact
R7s

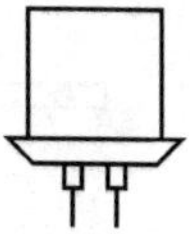

PG12

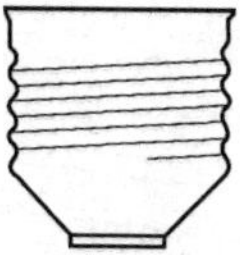

Medium screw
E26

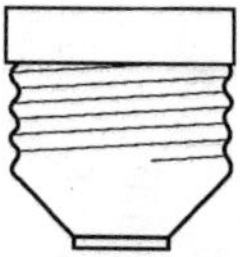

Export screw
E27

COMMON HID LAMP BASES WITH ANSI DESIGNATIONS *(cont.)*

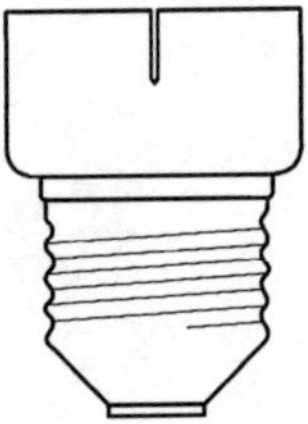

Skirted Ad medium
D8

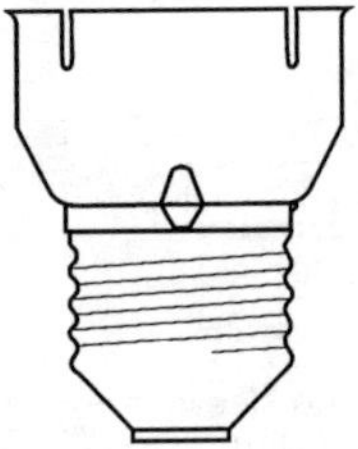

Medium skirt
E26/50x39

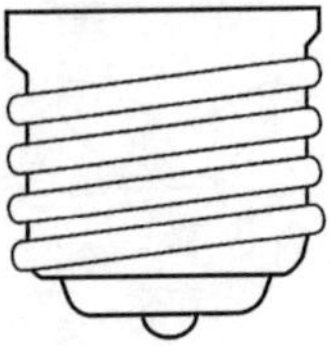

Mogul screw
E39

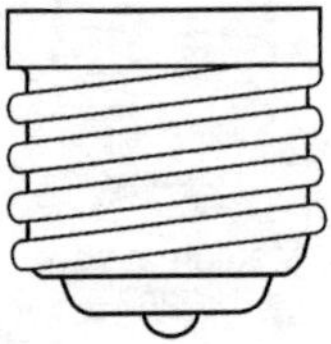

Export mogul
E40

COMMON HID LAMP BASES WITH ANSI DESIGNATIONS *(cont.)*

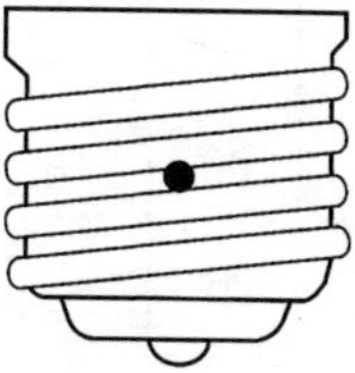

Position
oriented mogul
EP39

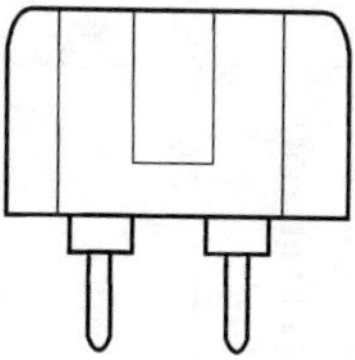

Bipin
G12

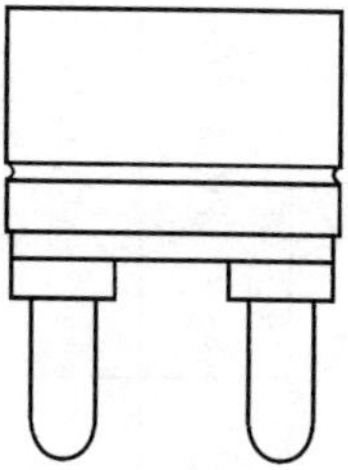

Mogul bipost
G38

Extended mogul
end prong
GX16d

HID BALLAST SELECTION FACTORS

Factor	Non-Regulating (reactor, lag)	Lead-Type Regulator (CWA)	Lag-Type Regulator (magnetic)
Line Voltage Variation	+/- 5%	+/- 10%	+/- 10%
Losses	low	medium to high	high
Power Factor	40 to 50%	90% +	90% +
Dip Tolerance	20 to 10%	50 to 10%	60 to 30%
Lamp Wattage Regulation	2 to 2½% for each 1% change of line voltage	1 to 1½% for each 1% change of line voltage	¾% for each 1% change of line voltage

BALLAST LOSS

Lamp	Lamp Rated Wattage (in W)	Ballast Loss (power loss %)
Low-pressure sodium	70	27
	100	25
	150	22
	250	20
	400	15
	1000	7
Mercury-vapor	40	32
	75	25
	100	22
	175	17
	250	16
	400	14
	700	7
	1000	7
Metal-halide	175	20
	250	17
	400	13
	1000	7
	1500	7
High-pressure sodium	50	30
	100	24
	150	22
	250	20
	400	16
	1000	7

RECOMMENDED BALLAST OUTPUT VOLTAGE LIMITS			
Ballast	Lamp Size		rms Voltage (Volts)
	Wattage	ANSI Number	
Low-pressure sodium	18	L69	300-325
	35	L70	455-505
	55	L71	455-505
	90	L72	455-525
	135	L73	645-715
	180	L74	645-715
Mercury-vapor	50	H46	225-255
	75	H43	225-255
	100	H38	225-255
	175	H39	225-255
	250	H37	225-255
	400	H33	225-255
	700	H35	405-455
	1000	H36	405-455
Metal-halide	70	M85	210-250
	100	M90	250-300
	150	M81	220-260
	175	M57	285-320
	250	M80	230-270
	250	M58	285-320
	400	M59	285-320
	1000	M47	400-445
	1500	M48	400-445
High-pressure sodium	35	S76	110-130
	50	S68	110-130
	70	S62	110-130
	100	S54	110-130
	150	S55	110-130
	150	S56	200-250
	200	S66	200-230
	250	S50	175-225
	310	S67	155-190
	400	S51	175-225
	1000	S52	420-480

RECOMMENDED SHORT-CIRCUIT CURRENT TEST LIMITS

Ballast	Lamp Size		Short-Circuit Current Range (Amps)
	Wattage	ANSI Number	
Low-pressure sodium	18	L69	.30 - .40
	35	L70	.52 - .78
	55	L71	.52 - .78
	90	L72	.8 - 1.2
	135	L73	.8 - 1.2
	180	L74	.8 - 1.2
Mercury-vapor	50	H46	.85 - 1.15
	75	H43	.95 - 1.70
	100	H38	1.10 - 2.00
	175	H39	2.0 - 3.6
	250	H37	3.0 - 3.8
	400	H33	4.4 - 7.9
	700	H35	3.9 - 5.85
	1000	H36	5.7 - 9.0
Metal-halide	70	M85	.85 - 1.30
	100	M90	1.15 - 1.76
	150	M81	1.75 - 2.60
	175	M57	1.5 - 1.90
	250	M80	2.9 - 4.3
	250	M58	2.2 - 2.85
	400	M59	3.5 - 4.5
	1000	M47	4.8 - 6.15
	1500	M48	7.4 - 9.6
High-pressure sodium	35	S76	.85 - 1.45
	50	S68	1.5 - 2.3
	70	S62	1.6 - 2.9
	100	S54	2.45 - 3.8
	150	S55	3.5 - 5.4
	150	S56	2.0 - 3.0
	200	S66	2.50 - 3.7
	250	S50	3.0 - 5.3
	310	S67	3.8 - 5.7
	400	S51	5.0 - 7.6
	1000	S52	5.5 - 8.1

BALLAST SOUND RATING

Sound Rating	Noise Level (decibels)	Recommended Application
A	20-24	Broadcasting booths, churches, study areas
B	25-30	Libraries, classrooms, quiet office areas
C	31-36	General office areas, commercial buildings
D	37-42	Retail stores, stock rooms
E	43-48	Light production areas, general outdoor lighting
F	Over 48	Street lighting, heavy production areas

TYPICAL RADIO INTERFERENCE FILTER FOR HID BALLAST

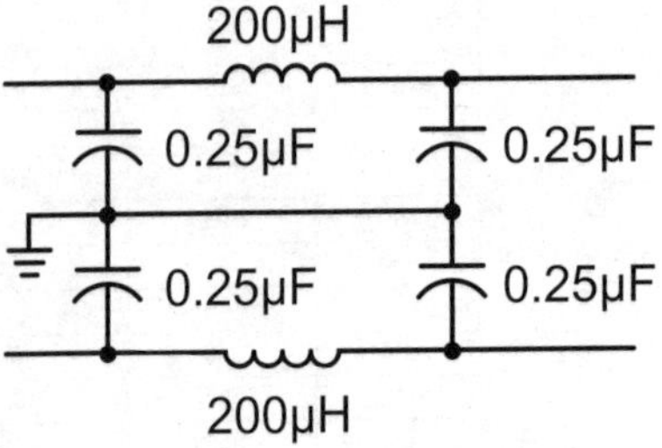

HID LAMP TROUBLESHOOTING GUIDE

Problem	Possible Cause	Corrective Action
Lamp does not start	Normal end of life operating characteristic	Replace with new lamp.
	Loose lamp connection	Re-seat lamp in socket securely. Ensure lamp holder is rigidly mounted and properly spaced.
	Defective photocell used for automatic turn ON	Replace defective photocell.
	Low line voltage applied to circuit	Ensure that voltage is within ±7% of rated voltage.
	Cold area or draft hitting lamp	Protect lamp with enclosure. Use a low temperature-rated ballast.
	Defective ballast	Replace ballast.
Lamp cycles ON and OFF or flickers when ON	Normal end of life operating characteristic	Replace with new lamp.
	Poor electrical connection or loose bulb	Check electrical connections and socket contacts.
	Line voltage variations	Ensure that voltage is within ±7% of rated voltage. Move lamps to separate circuit when lamps are on same circuit as high-power loads. High overloads cause a voltage dip when turned ON. This voltage dip may cause the lamp to turn OFF.
Short lamp life for line voltage	Wrong wattage lamp or ballast	HID lamps and ballast must be matched in size. Lamp life is shortened when a large-wattage lamp is used. The same size and type ballast must be installed when replacing a ballast.
	Power tap set too low for the correct supply voltage	Check ballast taps and ensure they are set for the correct supply voltage.
	Defective sodium ballast starter	Replace with new lamp.
Low light output	Normal end of life operating characteristic	Replace with new lamp.
	Dirty lamp or lamp fixture	Keep lamp fixture clean.
	Early blackening of bulb caused by incorrect lamp size or ballast	Ensure that lamp size, lamp type, and ballast match.

OUTDOOR LUMINAIRES

Luminaire	Illuminance Distribution
Type I	Narrow, symmetric illuminance pattern
Type II	Slightly wider illuminance pattern than Type I
Type III	Wide illuminance pattern
Type IV	Widest illuminance pattern
Type V	Symmetrical circular illuminance pattern
Type VI	Symmetrical, nearly square illuminance pattern

CUTOFF CLASSIFICATIONS

Type	Luminaire Intensity Distribution
Full cutoff	Where zero candela intensity occurs at an angle of 90° above nadir and at all greater angles from nadir. Candela per 1000 lamp lumens does not exceed 100 (10%) at a vertical angle of 80° above nadir. Applies to all lateral angles around luminaire.
Cutoff	Where the candela per 1000 lamp lumens does not numerically exceed 25 (2.5%) at an angle of 90° above nadir, and 100 (10%) at a vertical angle of 80° above nadir. Applies to all lateral angles around luminaire.
Semicutoff	Where the candela per 1000 lamp lumens does not exceed 50 (5%) at an angle of 90° above nadir and 200 (20%) at a vertical angle of 80° above nadir. Applies to all lateral angles around luminaire.
Noncutoff	Where there is no candela limitation in the zone above maximum candela.

GENERAL MOUNTING GUIDELINES FOR HID LAMP POLES

LH = LAMP HEIGHT

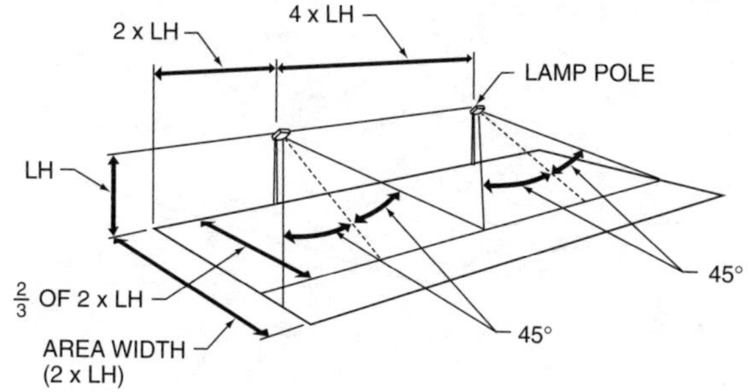

AIMED ILLUMINATION

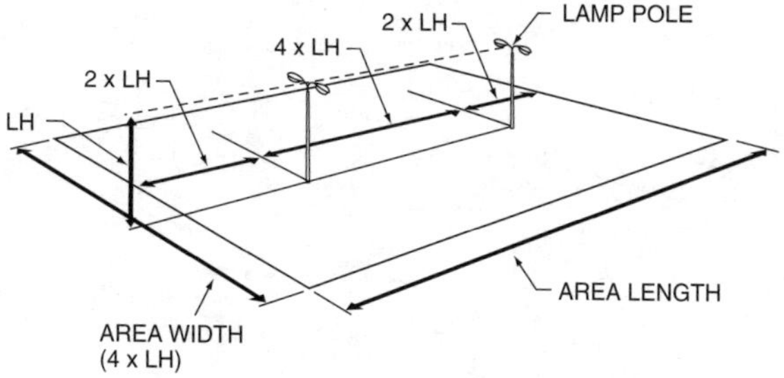

GENERAL ILLUMINATION

GENERAL MOUNTING GUIDELINES FOR HID LAMP POLES *(cont.)*

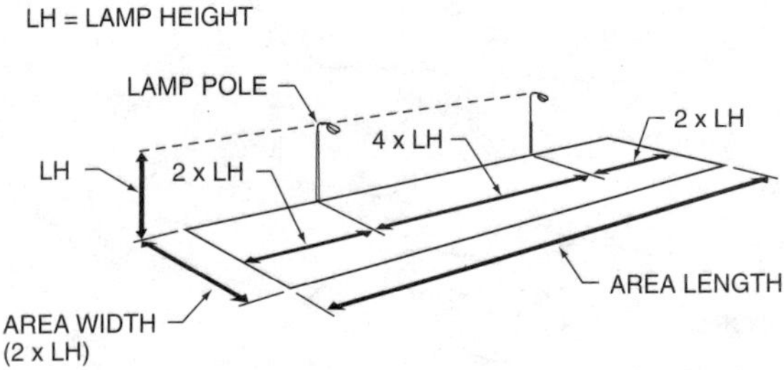

ILLUMINATION FROM ONE SIDE

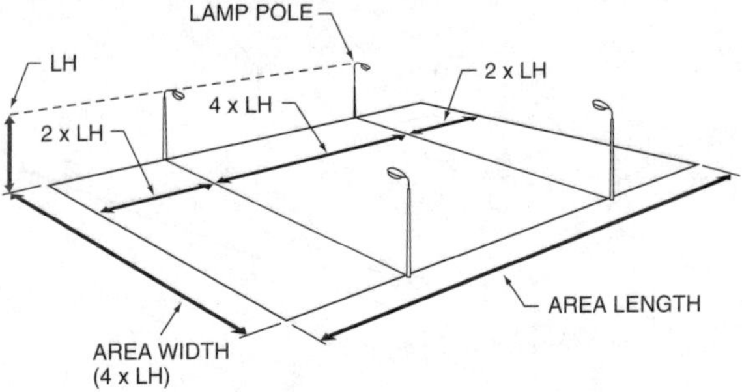

ILLUMINATION FROM TWO SIDES

LOW-PRESSURE SODIUM

Although low-pressure sodium (LPS) lamps are similar to fluorescent systems (both are low pressure), they are commonly included in the HID family. LPS lamps are the most efficacious light sources (that is, they put out the most light per watt of electricity), but produce the poorest quality light of all lamp types. LPS lamps are *monochromatic* light sources, putting out only one wavelength of light. When illuminated by LPS lamps, all colors appear as black, white or shades of gray. LPS lamps are available in wattages ranging from 18-180.

LPS lamp use has been generally limited to outdoor applications such as security or street lighting and indoor, low-wattage applications where color quality is not important (stairwells). However, because the color rendering of LPS lamps is so poor, many municipalities do not allow them for roadway lighting. Because LPS lamps are long tubes (like thin fluorescent lamps), they are less effective in directing and controlling a light beam, compared with *point sources* like high-pressure sodium and metal-halide. Therefore, lower mounting heights will provide better results. To compare LPS installations with other alternatives, calculate the installation efficacy as the average maintained foot-candles divided by the input watts per square foot of illuminated area.

The input wattage of an LPS system increases over time, and maintains consistent light output over the lamp life. One hazard: Low-pressure sodium lamps can explode if the sodium comes in contact with water. You should always dispose of these lamps according to manufacturer's instructions.

LOW-PRESSURE SODIUM BALLASTS

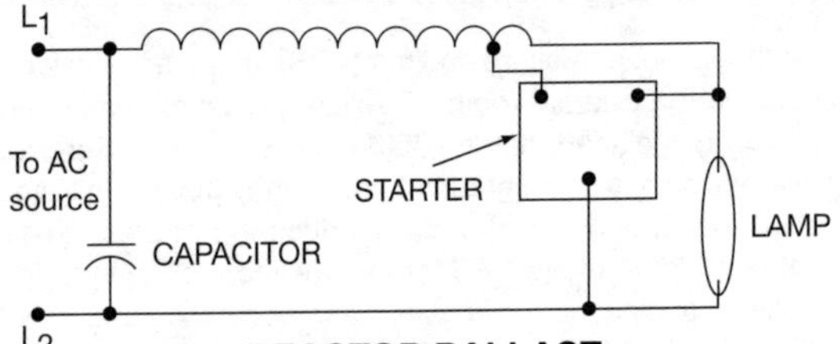

REACTOR BALLAST

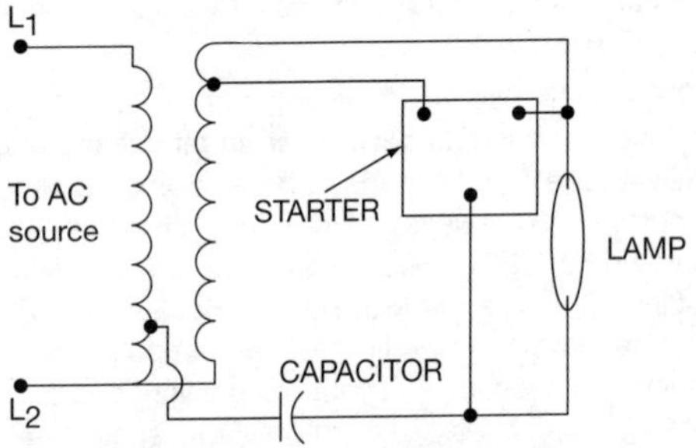

AUTOREGULATOR BALLAST

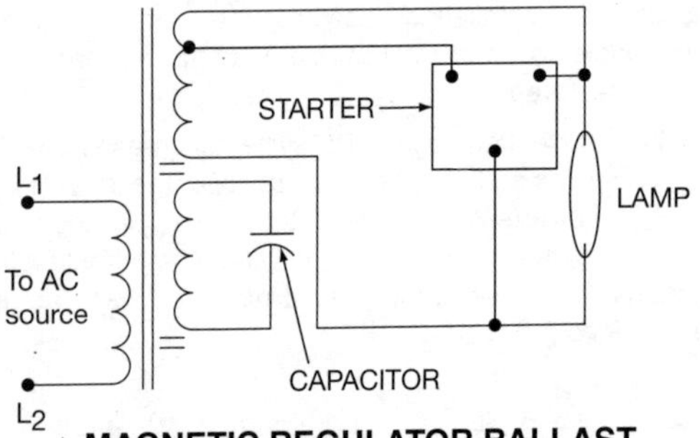

MAGNETIC REGULATOR BALLAST

LOW-PRESSURE SODIUM LAMP FIXTURES

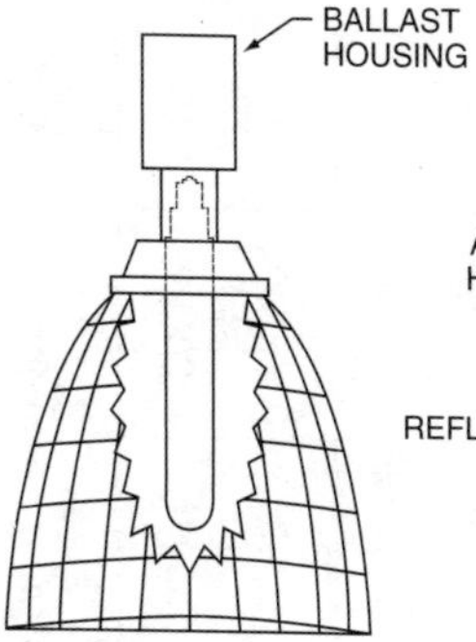

HIGH-PROFILE

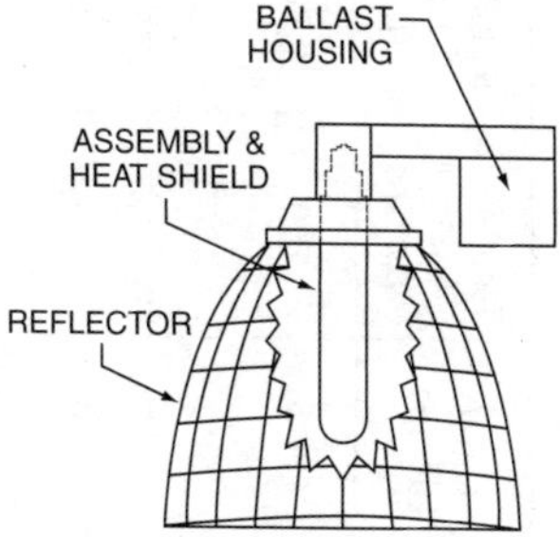

LOW-PROFILE

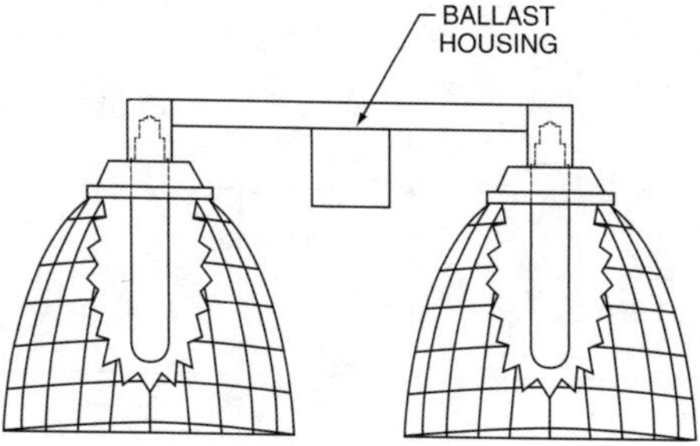

TWIN OPTICAL

TYPICAL PERFORMANCE VALUES FOR LOW-PRESSURE SODIUM SYSTEMS

Lamp Types	Lamp Watts	System Watts	Initial Lamp CRI	Lamp Lumens	Maintained System Lumens	Maintained System Efficacy	Rated Life/Hrs
8-inch	18	36	0	1800	1800	50	14000
12-inch	35	60	0	4800	4800	80	18000
17-inch	55	80	0	8000	8000	100	18000
21-inch	90	125	0	13500	13500	108	18000
30-inch	135	178	0	22500	22500	126	18000
44-inch	180	220	0	33000	33000	150	18000

NOTE: Maintained performance includes effect of lamp lumen depreciation (@40% rated life). Lamp life ratings are based on 10 hours per start.

LPS WATTS RISE OVER LIFE HOURS

Lamp Types	Lamp Watts	Life Hours 100	2000	5000	10000	18000	Average Watts Over Life
8-inch	18	17	18	20	18	–	18.6
12-inch	35	35	36	37	38	39	37.5
17-inch	55	55	56	58	60	61	58.9
21-inch	90	90	95	97	97	98	96.3
30-inch	135	130	134	140	141	142	139.1
44-inch	180	176	182	190	191	192	188.9

Representative watts rise over life characteristics for lamps with heat insulating shields at the end of the arc tube.

MORTALITY CURVES FOR TYPICAL LOW-PRESSURE SODIUM LAMPS

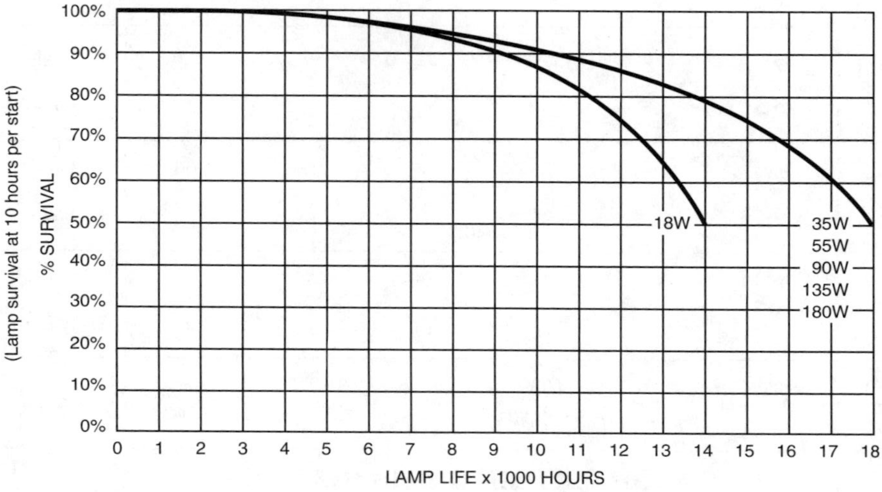

SPECTRAL POWER DISTRIBUTION FOR LOW-PRESSURE SODIUM LAMPS

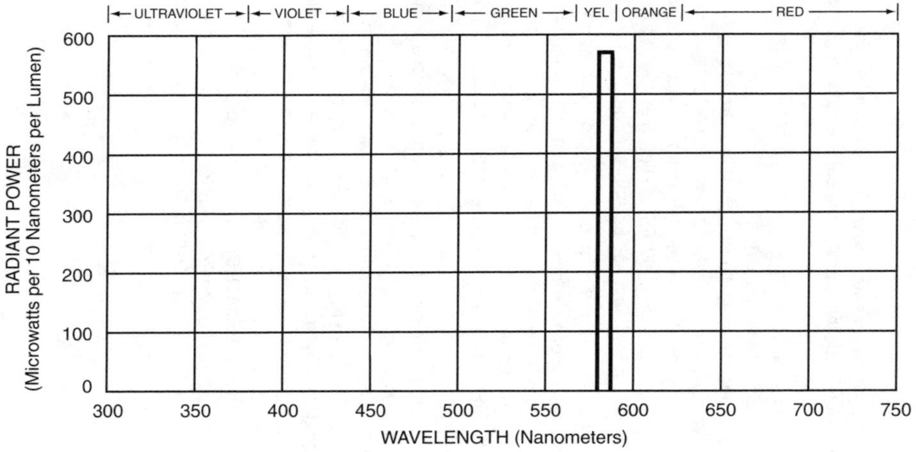

HIGH-PRESSURE SODIUM

The high-pressure sodium (HPS) lamp is widely used for outdoor and industrial applications. Its higher efficacy makes it a better choice than metal-halide for many applications, especially when good color rendering is not a priority. HPS lamps differ from mercury and metal-halide lamps in that they do not contain starting electrodes; the ballast circuit includes a high-voltage electronic starter. The arc tube is made of a ceramic material which can withstand temperatures over 2300°F. It is filled with xenon to help start the arc, as well as a sodium-mercury gas mixture.

The efficacy of the lamp is high (as much as 140 lumens per watt). For example, a 400-watt high-pressure sodium lamp produces 50,000 initial lumens. The same wattage metal-halide lamp produces 40,000 initial lumens, and the 400-watt mercury-vapor lamp produces only 21,000 initial lumens.

Sodium, the major element used, produces a gold color that is characteristic of HPS lamps. Although HPS lamps are not generally recommended for applications where color rendering is critical, HPS color rendering properties are being improved. Some HPS lamps are now available in "deluxe" and "white" colors that provide higher color temperature and improved color rendition. The efficacy of low-wattage "white" HPS lamps is lower than that of metal-halide lamps (lumens per watt of low-wattage metal-halide is 75-85, while white HPS is 50-60 lumens per watt).

HIGH-PRESSURE SODIUM BALLASTS

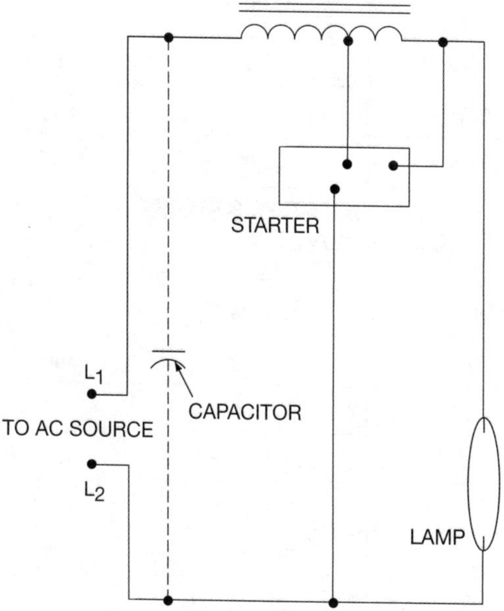

250 W HIGH-PRESSURE SODIUM LAMP CURRENTS

Lamp Fixture Voltage	Starting Current	Operating Current
120	1.60	2.70
208	.95	1.50
240	.85	1.36
277	.70	1.15
480	.40	.65

HIGH-PRESSURE SODIUM LAMP LUMENS DATA

Lamp Wattage	Initial Lumens	Lumens per Watt
50	3400	68
70	5600	80
100	7800	78
150	14,000	93
175	16,000	91
250	20,000	80
400	40,000	100
1000	117,000	117
1500	155,000	103
1650	177,000	107

TYPICAL PERFORMANCE VALUES FOR HIGH-PRESSURE SODIUM SYSTEMS

Lamp Types	Lamp Watts	System Watts	Lamp CRI	Initial Lamp Lumens	Maintained System Lumens	Maintained System Efficacy	Rated Life/Hrs
Standard	35	45	22	2250	2025	45	24000
Standard	50	65	22	4000	3600	55	24000
Standard	70	95	22	6400	5450	57	24000
Standard	100	130	22	9500	8550	66	24000
Standard	150	195	22	16000	14400	74	24000
Standard	250	300	22	28000	27000	90	24000
Standard	400	465	22	51000	45000	97	24000
Standard	1000	1100	22	140000	126000	115	24000
Energy Saver	225	275	22	27500	24800	90	24000
Energy Saver	360	425	22	45000	40500	95	24000
Deluxe	70	95	60	4400	3960	42	15000
Deluxe	100	130	60	7300	6570	51	15000
Deluxe	150	195	60	12000	10800	55	15000
Deluxe	250	300	65	23000	20700	69	15000
Deluxe	400	465	65	37500	33750	73	15000
White	35	45	70	1250	1000	22	10000
White	50	65	70	2300	1725	27	10000
White	100	130	70	4700	3520	27	10000
Retrofit for MV	150	195	25	15000	13500	69	24000
Retrofit for MV	215	265	25	23000	20700	78	24000
Retrofit for MV	360	425	25	45000	40500	95	24000

HIGH-PRESSURE SODIUM LAMP LUMEN DEPRECIATION

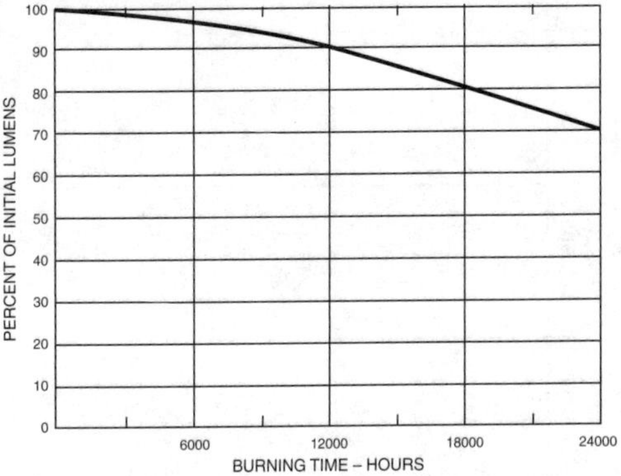

Lumen Maintenance of 150, 250, 400 and 1000 Watt Lumalux Lamps

MORTALITY OF TYPICAL HIGH-PRESSURE SODIUM LAMPS

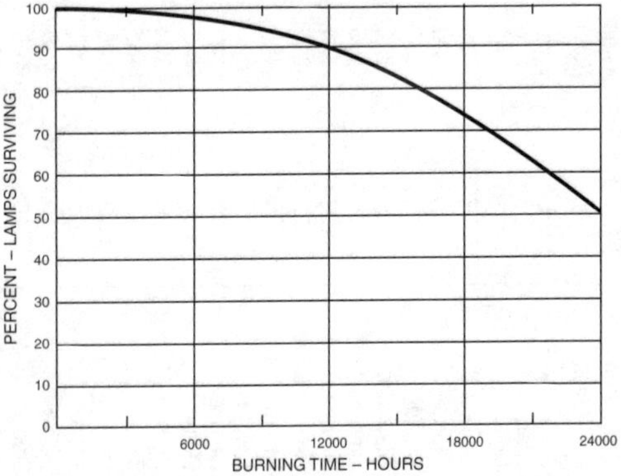

SPECTRAL POWER DISTRIBUTION FOR HIGH-PRESSURE SODIUM LAMPS

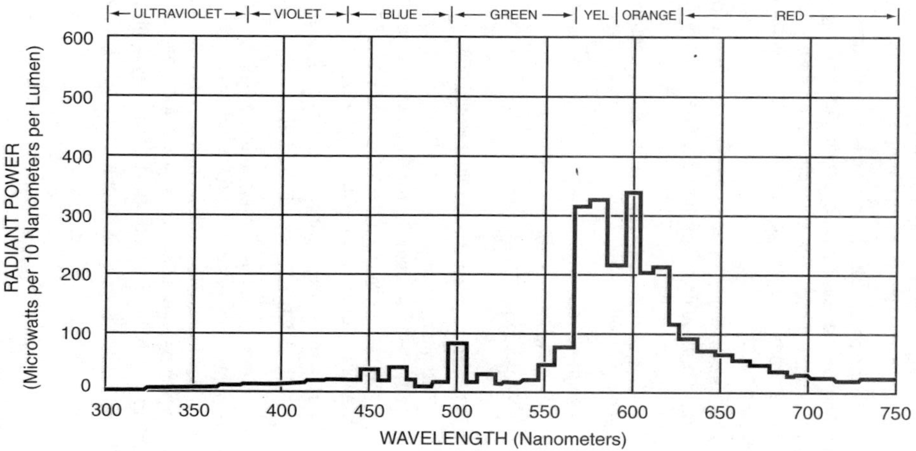

METAL-HALIDE

These lamps are similar to mercury-vapor lamps but use metal-halide additives inside the arc tube along with the mercury and argon. These additives enable the lamp to produce more visible light per watt, along with improved color rendering.

Wattages range from 32 to 2,000, offering a wide range of indoor and outdoor applications. The efficacy of metal-halide lamps ranges from 50 to 115 lumens per watt (about double that of mercury-vapor). The advantages of metal-halide lamps are:

- High efficacy
- Good color rendering
- Wide range of wattages

But, they also have some operating limitations:

- The rated life of metal-halide lamps is shorter than other HID sources (low-wattage lamps last less than 7500 hours while high-wattage lamps last an average of 15,000 to 20,000 hours)
- The color of light varies, and may shift over the life of the lamp

Because of good color rendition and a high lumen output, these lamps are good for sports arenas and stadiums. Indoor uses include large auditoriums and convention halls. These lamps are sometimes used for general outdoor lighting, such as parking facilities, but a high-pressure sodium system is more popular for such uses.

METAL-HALIDE BALLASTS

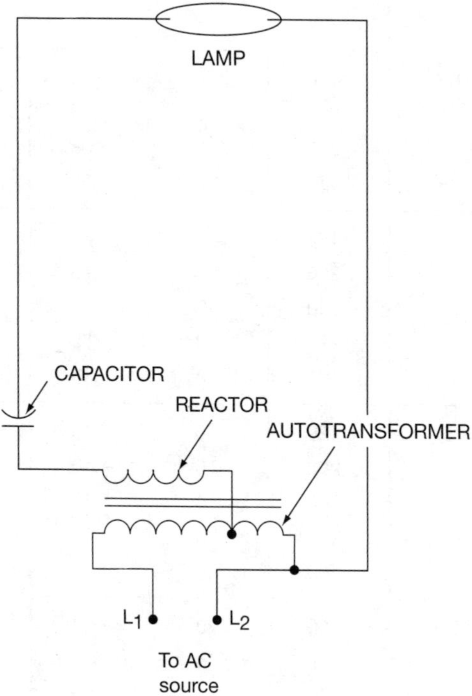

METAL-HALIDE LAMP CHARACTERISTICS				
Lamp Wattage (W)	Voltage Rating (V)	Fuse Rating (A)	Starting Current (A)	Operational Current (A)
50	120	3	.60	.65
	277	3	.25	.30
100	120	8	1.15	1.15
	208	5	.66	.66
	240	3	.58	.58
	277	3	.50	.50
175	120	5	1.30	1.80
	208	3	.75	1.05
	240	3	.65	.90
	277	3	.55	.80
	480	3	.35	.45
250	120	8	2.10	2.50
	208	5	1.40	1.45
	240	5	1.10	1.25
	277	3	1.00	1.10
	480	3	.60	.60
1000	120	20	8.0	9.0
	208	15	4.6	5.2
	240	10	4.0	4.5
	277	10	3.5	3.9
	480	10	2.0	2.3

TYPICAL PERFORMANCE VALUES FOR METAL-HALIDE SYSTEMS

Lamp Types	Lamp Watts	System Watts	Lamp CRI	Initial Lamp Lumens	Maintained System Lumens	Maintained System Efficacy	Rated Life/Hrs
2⅛" Diameter	50	67	65	3500	2550	38	5000
2⅛" Diameter	70	90	65	5500	4000	44	7500
2⅛" Diameter	100	127	65	9000	6390	50	15000
2⅛" Diameter	150	195	65	13500	10200	52	15000
2⅞" Diameter (1)	175	210	65	15000	12000	57	10000
3½" Diameter (1)	250	293	65	23000	18000	61	10000
4⅝" Diameter (1)	400	458	65	40000	32000	70	20000
7" Diameter (1)	1000	1080	65	115000	92000	85	12000
7" Diameter (1)	1500	1620	65	155000	140000	86	6000
Retrofit for MV	325	383	65	28000	18200	48	20000
Retrofit for MV	950	1030	65	100000	80000	78	12000
Retrofit for HPS	250	300	65	18000	13500	45	10000
Retrofit for HPS	400	465	65	40000	30000	65	20000
Energy Saver MH	150	185	65	13500	10200	55	10000
Energy Saver MH	225	268	65	19000	14300	53	10000
Energy Saver MH	360	418	65	35000	26300	63	10000
Super MH (2)	150	–	65	15000	11300	–	15000
Super MH (2)	200	–	65	21000	15800	–	15000
Super MH (2)	350	375	65	36000	27000	72	15000

(1) High-output lamps designed for specific orientation and produce 10-25% more light (2) Requires external igniter in ballast.

LAMP LUMEN DEPRECIATION FOR TYPICAL 400-WATT METAL-HALIDE LAMPS

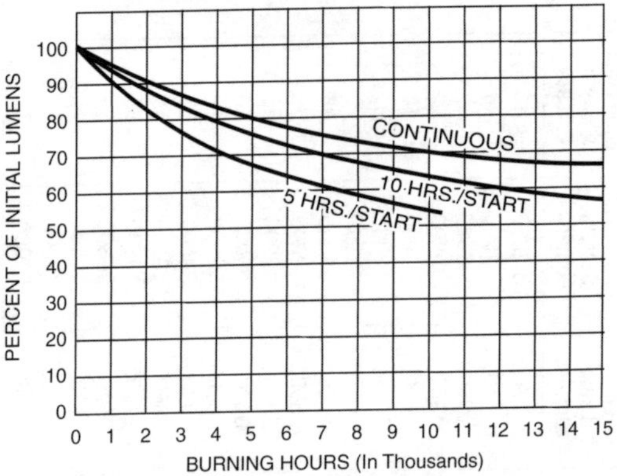

MORTALITY CURVES FOR TYPICAL 400-WATT METAL-HALIDE LAMPS

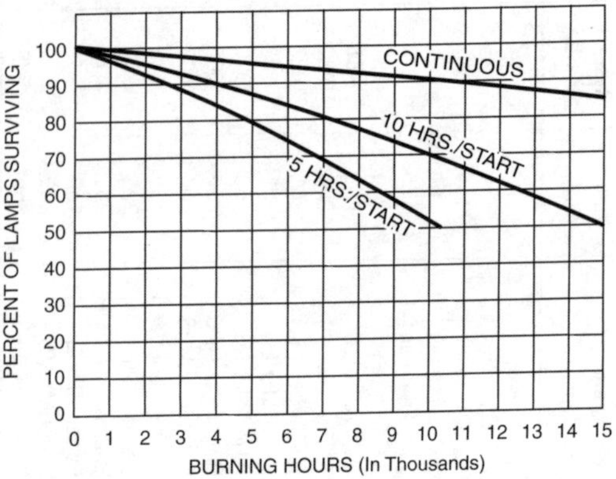

SPECTRAL POWER DISTRIBUTIONS FOR CLEAR METAL-HALIDE LAMPS

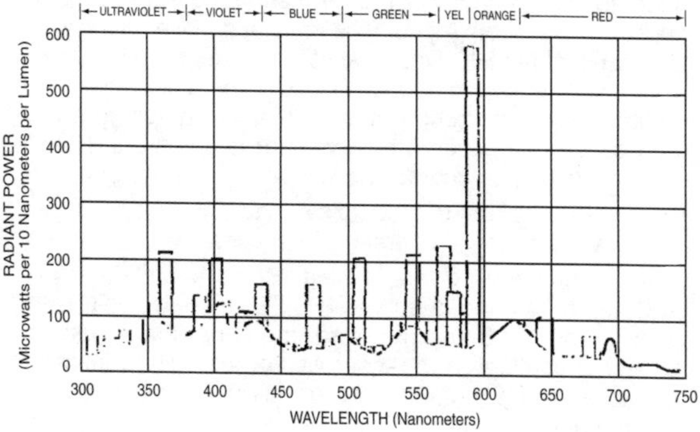

SPECTRAL POWER DISTRIBUTIONS FOR PHOSPHOR-COATED METAL-HALIDE LAMPS

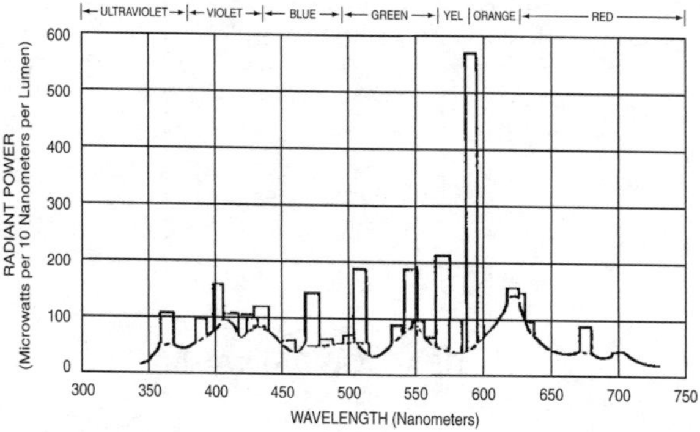

MERCURY-VAPOR

Clear mercury-vapor lamps produce a white light with blue tones. They consist of an arc tube with tungsten electrodes at both ends. These lamps have the lowest efficacies of the HID types, rapid lumen depreciation and a low color rendering index. Because of this, other HID sources have replaced mercury-vapor lamps in most applications. However, mercury-vapor lamps are still used due to two of their characteristics:

- Long life - 24,000 hours average
- Vivid portrayal of green landscapes

The mercury-vapor arc is filled with high purity mercury and argon gas and is enclosed within an outer bulb filled with nitrogen. Color-improved mercury lamps use a phosphor coating on the inner wall of the bulb to improve the color rendering index resulting in slight reductions in efficiency.

MERCURY-VAPOR BALLASTS

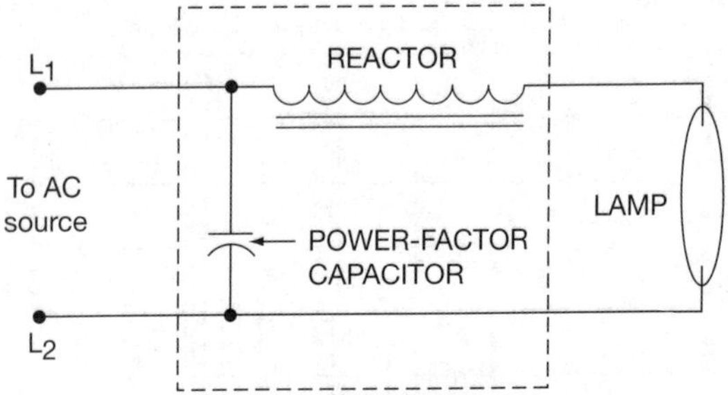

REACTOR BALLAST

MERCURY-VAPOR BALLASTS (*cont.*)

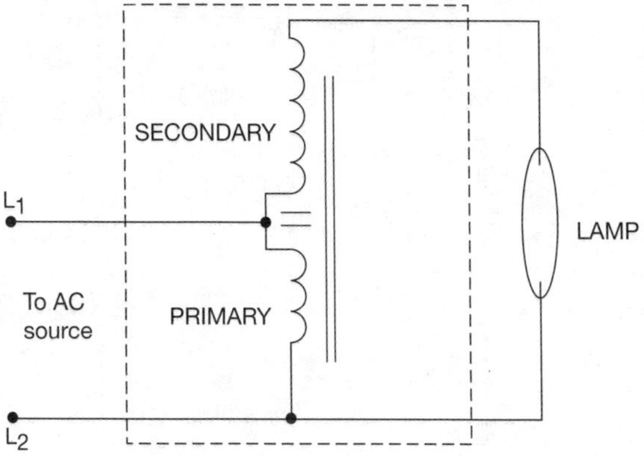

LAG BALLAST

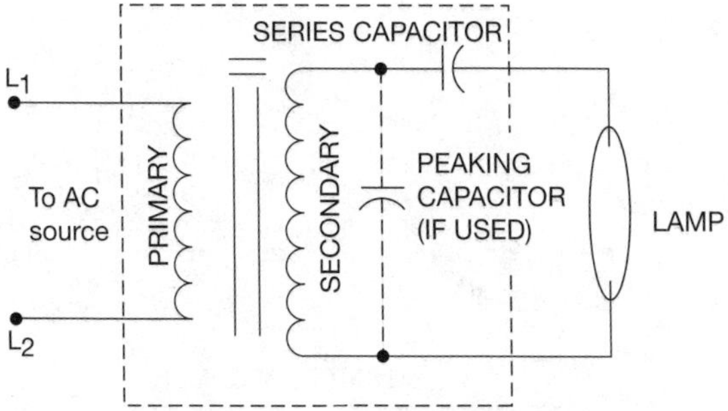

REGULATOR BALLAST

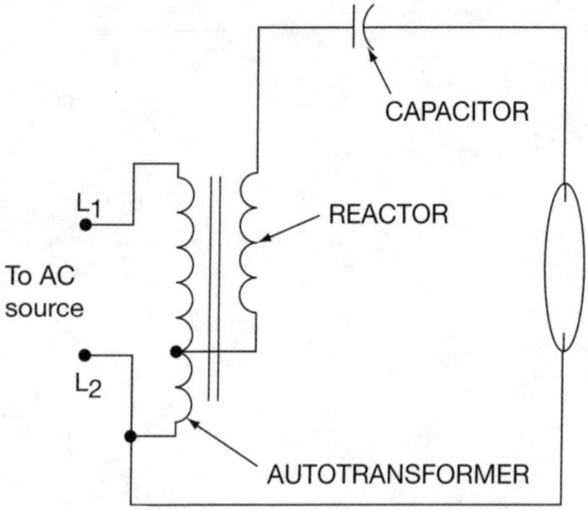

AUTOTRANSFORMER (CONSTANT-WATTAGE)

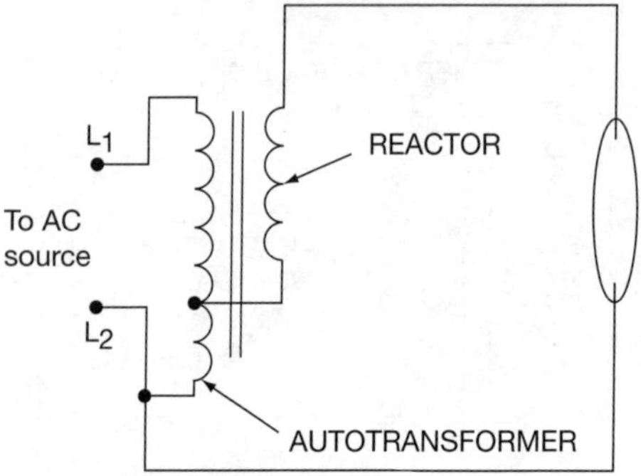

AUTOTRANSFORMER (HIGH-REACTANCE)

TYPICAL PERFORMANCE VALUES FOR MERCURY-VAPOR SYSTEMS

Lamp Types	Lamp Watts	System Watts	Lamp CRI	Initial Lamp Lumens	Maintained System Lumens	Maintained System Efficacy	Rated Life/Hrs
2⅛" Diameter (Coated)	50	67	50	1575	1250	19	16000
2⅛" Diameter (Coated)	75	92	50	2800	2250	24	16000
3" Diameter	100	120	15	4100	3450	29	24000
3½" Diameter	175	206	15	7900	7400	36	24000
3½" Diameter	250	284	15	12100	10500	37	24000
4⅝" Diameter	400	458	15	21000	18900	41	24000
7" Diameter	1000	1050	15	60000	45000	43	24000

APPROX. LUMEN MAINTENANCE OF COLOR-IMPROVED MERCURY LAMPS OPERATING VERTICALLY

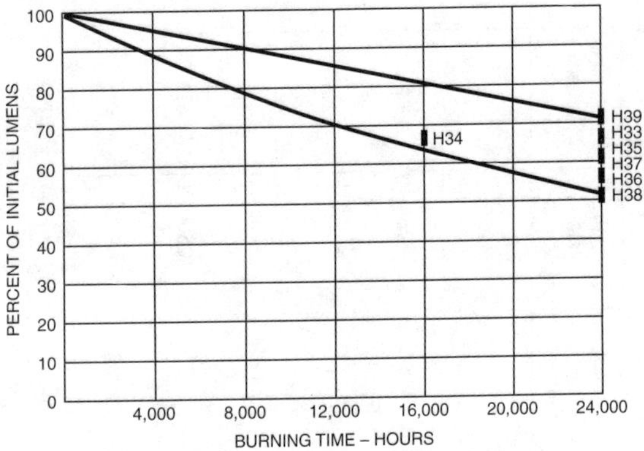

MORTALITY CURVES FOR TYPICAL MERCURY-VAPOR LAMPS

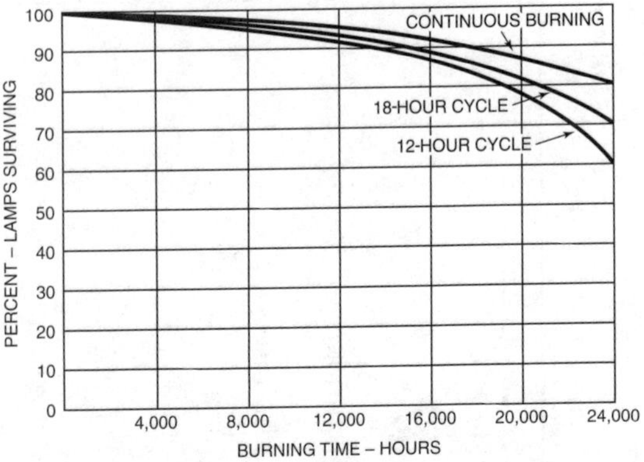

SPECTRAL POWER DISTRIBUTION FOR CLEAR MERCURY-VAPOR LAMPS

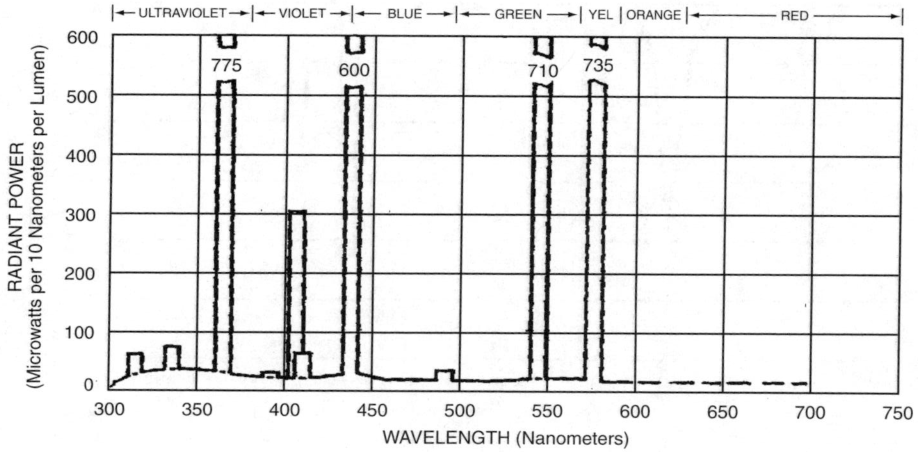

SPECTRAL POWER DISTRIBUTION FOR DELUXE MERCURY-VAPOR LAMPS

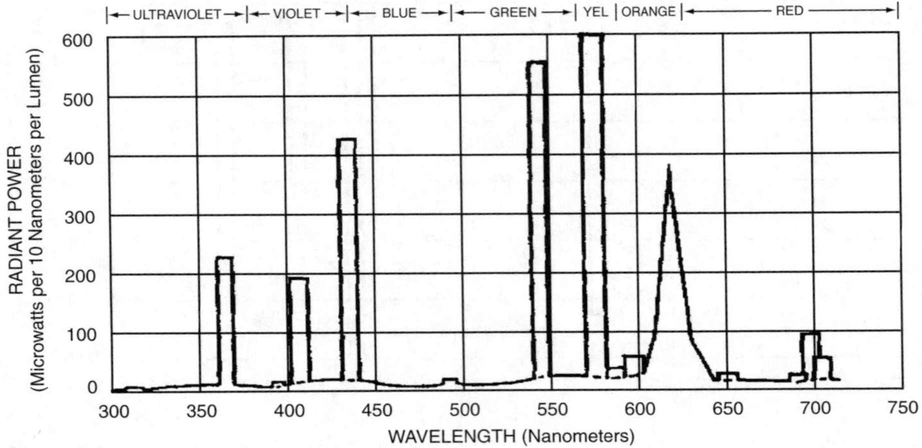

CHAPTER 3
FLUORESCENTS

Fluorescent lamps are the most commonly used commercial light source because of their relatively high efficacy, diffuse light distribution characteristics, and long operating life. Full-size lamps are available in straight, U-shaped, and circular configurations. Lamp diameters range from 1" to 2.5". The most common lamp type is the four-foot (F40), 1.5" diameter (T12) straight fluorescent lamp. More efficient lamps are now available in smaller diameters, including the T10 (1.25") and T8 (1").

Fluorescent lamps are available in color temperatures ranging from warm (2700K) "incandescent-like" to very cool (6500K) "daylight." "Cool white" (4100K) is the most common. "Neutral white" (3500K) is becoming popular for office and retail use. With fluorescent lamp-ballast systems, light output, input watts, and efficacy are sensitive to changes in ambient temperature. When it is significantly above or below 25°C (77°F), the performance of the system can change.

Compact fluorescent lamps are long-lasting, energy-efficient substitutes for incandescents. The wattages of compacts range from 5 to 40 (replacing incandescents ranging from 25 to 150 watts) and provide energy savings of 60 to 75 percent. While producing light similar in color to incandescents, the life expectancy of a compact fluorescent is about 10 times that of an incandescent lamp. The use of compacts is very limited in dimming applications, and at unusual ambient temperatures. They may also cause interference with some electronic equipment.

Compact fluorescents are covered in detail in Chapter 4, where they are compared to the incandescent lamps they were designed to replace.

FLUORESCENT LAMP IDENTIFICATION

Fluorescent lamp codes consist of a series of letters and numbers.

Tubular preheat lamps

F4T5/CW	F	4	T	5	CW
	↑	↑	↑	↑	↑
	fluorescent	watts	shape (tubular)	diameter in ⅛"	color

Tubular instant-start lamps

F96T12/CW	F	96	T	12	CW
	↑	↑	↑	↑	↑
	fluorescent	length* (inches)	shape (tubular)	diameter in ⅛"	color

**except F40T12/IS and F40T17/IS, where the first number represents watts, and "IS" denotes instant start.*

Tubular rapid-start lamps 40 watts and less

F30T12/RS/CW	F	30	T*	12*	RS*	CW
	↑	↑	↑	↑	↑	↑
	fluorescent	watts	shape	diameter in ⅛"	rapid start	color

**omitted for standard 40-watt lamp*

Tubular 800-milliamp, rapid-start lamps

F96T12/HO/CW	F	96	T	12	HO	CW
	↑	↑	↑	↑	↑	↑
	fluorescent	length (inches)	shape	diameter in ⅛"	high output	color

Tubular 1500-milliamp, rapid-start lamps

F48T12/CW/1500	F	48	T	12	CW	1500
	↑	↑	↑	↑	↑	↑
	fluorescent	length (inches)	shape	diameter in ⅛"	color	current

COMMON FLUORESCENT TYPES

Type	Watts	Length	Size	Base
T-5	4 – 13	6 – 21	⅝	Miniature Bi-pin
T-5	13	21	⅝	Miniature Bi-pin
T-8	15 – 30	18 – 36	1	Medium Bi-pin
T-9	20 – 40	6 – 16	1⅛	Four-pin
T-10	40	48	1¼	Medium Bi-pin
T-12	14 – 75	15 – 96	1½	Medium Bi-pin
T-12	60 – 75	96	1½	Single-Pin

Note: All lengths and sizes are in inches.

COMMON SHAPE IDENTIFIERS

C – Circular lamp. The letter "C" typically appears in the prefix, immediately after the "F."

T – Tubular lamp, omitted from the F40 lamp code for rapid-start lamps.

J – Used in conjunction with a "T" designator, "J" denotes a lamp with a clear outer glass jacket over the regular "T" envelope, which acts as a thermal insulator and permits operation of the lamps at sub-zero temperatures.

BX, **PL**, and **PTT**
– These identifiers denote a family of single ended compact lamps.

U – A lamp shaped like a letter "U."

FLUORESCENT LAMP DATA

Lamp Type	Base	Watts	Lumens	Life/Hrs
T-12 (Preheat)	Med bi-pin	20	1250	9000
T-12/HO (High output)	RDC	35	1650	9000
T8 (Octron)	Med bi-pin	32	2850	20,000
T-12 (Rapid start)	Med bi-pin	34	2650	20,000
T-12 (Rapid start)	Med bi-pin	40	3150	20,000
T-12 (Slimline)	Single pin	40	3000	9000
BX (Long twin tube)	Single end	39	3150	20,000
T-12/HO (High output)	RDC	60	4300	12,000
T-12/VHO (Very high output)	RDC	110	6400	10,000
T-12 (Slimline)	Single pin	75	6150	12,000

FLUORESCENT LAMP OUTPUT

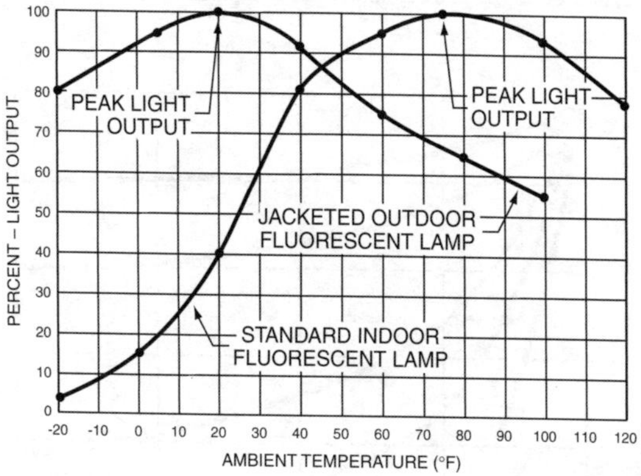

MORTALITY CURVE FOR FLUORESCENT LAMPS

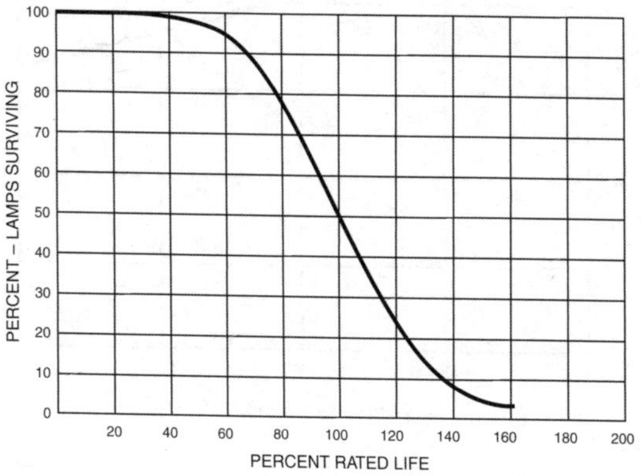

FLUORESCENT LAMP LUMEN DEPRECIATION

(3 hour per start)

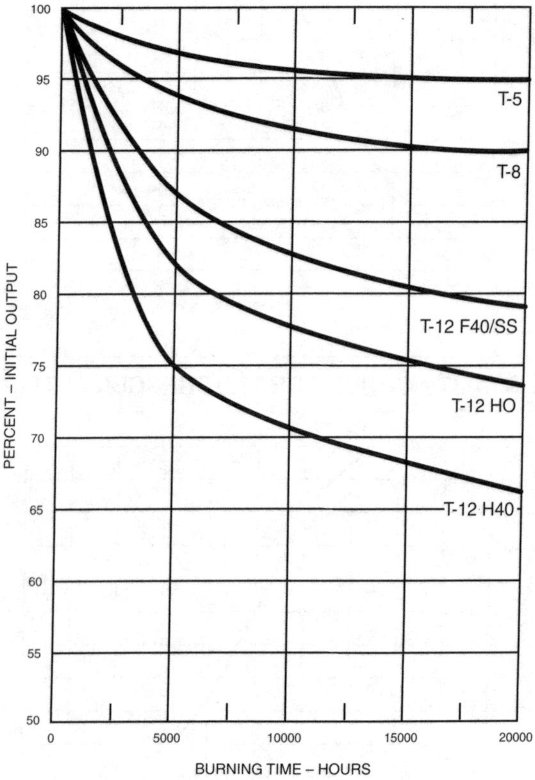

FLUORESCENT LAMP CHARACTERISTICS

Lamp Type & Description	Bulb Diameter (in.)	Bulb Watts	Base (End Caps)
F32T8	1	32	Med. Bi-pin
FT36W/2G11RS	⅝	38.1	2G11
CFL26	½	26	G24d-3
T5	⅝	28	Min. Bi-pin
T5/HO	⅝	54	2G11
40T12/SS	1	34-35	Med. Bi-pin
96T12/SS (Slimline)	1	60	Single pin
96T12/HO/SS (Rapid start)	1	97	Recess D.C.

FLUORESCENT LAMP CHARACTERISTICS *(cont.)*

Lamp Type & Description	Normal Length (mm)	(in.)	Min. Req. RMS Voltage (V) Reliable Starting	Operating Current (mA)
F32T8	1200	47.25	200	265
FT36W/2G11RS	419	16.49	1 = 230	430
CFL26	169	6.65	198	325
T5	1200	47.25	375	210
T5/HO	830	32.6	425	400
40T12/SS	1200	47.25	1 = 200 2 = 256	460
96T12/SS (Slimline)	2400	94.48	565	440
96T12/HO/SS (Rapid start)	2400	94.48	2 = 296	830

FLUORESCENT LAMP CHARACTERISTICS *(cont.)*			
Lamp Type & Description	Operating Voltage (V)	Cathode Heaters (Low Resistance)	
		Volts	Maximum Watts
F32T8	137	3.6	1.7
FT36W/2G11RS	110	3.6	continuous
CFL26	105	preheat >10 sec	420mA
T5	107	preheat <2 sec	210mA
T5/HO	135	preheat <2 sec	700mA
40T12/SS	79	3.6	–
96T12/SS (Slimline)	157	none	continuous
96T12/HO/SS (Rapid start)	126	3.6	continuous

2-FOOT x 4-FOOT TROFFER SHIELDING MEDIA

Shielding Material	EFF. %	VCP %
Standard Clear Lens	60-70	50-75
Low-Glare Clear Lens	60-75	75-85
Deep Cell Parabolic Louver	50-65	75-95
Translucent Diffuser	40-60	40-50
White Metal Louver	35-45	65-85
Small Cell Parabolic Louver	35-45	99

U.S. EPACT FLUORESCENT LAMP STANDARD LEVELS

Lamp Type	Nominal Lamp Wattage	Min. CRI	Min. Aver. Lamp Efficacy (lumens/watts)
4-foot Med. Bi-Pin	35W 35W	69 45	75.0 75.0
2-foot U-Shaped	35W 35W	69 45	68.0 64.0
8-foot Slimline	65W 65W	69 45	80.0 80.0
8-foot High Output	100W 100W	69 45	80.0 80.0

COMPACT FLUORESCENT FIXTURE DATA

Lamp Quantity & Size	120 Volt NPF Fixt. Total Amps/Watts	120 Volt HPF Fixt. Total Amps/Watts	277 Volt HPF Fixt. Total Amps/Watts	Lumens per Watt	Equivalent Incandescent Wattage	Standard Color Temp. / CRI	Min. Start Temp. (°F)
2 x 9W	.36/25	.20/25	.13/32	67	75W	2700° K / 82	25°
2 x 13W	.60/34	.28/34	.17/42	69	120W	2700° K / 86	32°
2 x 18W	.70/47	.44/47	.18/49	69	150W	2700° K / 86	23°
2 x 26W	1.0/64	.63/64	.26/54	69	200W	2700° K / 86	23°
1 x 9W	.18/13	.10/13	.065/13	67	40W	2700° K / 82	25°
1 x 13W	.30/17	.14/17	.085/17	69	60W	2700° K / 82	32°
1 x 18W	.35/24	.22/24	.09/24	69	75W	2700° K / 86	23°
1 x 26W	.50/32	.32/32	.13/32	69	100W	2700° K / 86	23°

TYPICAL PERFORMANCE VALUES

22.5" T5 Compact Fluorescents

FB40T12 and FB31T8 U-lamps perform essentially the same as F40T12 and F32T8 straight lamps, respectively.

Ballast Types / Lamp Types	Lamps per Ballast	Lamp Watts	System Input Watts	Lamp CRI
Old Standard Magnetic				
2 – F20T12 (preheat)	2	20	50	62
Standard EE Magnetic				
2 – F20T12 (preheat)	2	20	46	62
2 – F17T8	2	17	43	75
	2	17	43	85
2 – FT40T5	2	40	91	82
Electronic Rapid Start				
2 – F17T8	2	17	37	75
	2	17	37	85
Partial	2	17	27	75
Output	2	17	27	85
3 – F17T8	3	17	52	75
	3	17	52	85
4 – F17T8	4	17	70	75
	4	17	70	85
2 – FT40T5	2	40	74	82
3 – FT40T5	3	40	106	82
Electronic Instant Start				
2 – F17T8	2	17	33	75
2 – F17T8	2	17	33	85
3 – F17T8	3	17	47	75
3 – F17T8	3	17	47	85
4 – F17T8	4	17	62	75
4 – F17T8	4	17	62	85
2 – FT40T5	2	40	71	82
3 – FT40T5	3	40	101	82

NOTE: Different manufacturers of 22.5" lamps provide rated wattages of 38, 39 or 40 watts.

FOR 2-FOOT SYSTEMS

24" T8 Straight Fluorescent Lamps

Refer to the 4-foot table (2-lamp and 3-lamp) for representative values for FB40T12 and FB31T8 lamps.

Initial Lamp Lumens	Lamp Lumen Deprec.	Ballast Factor	Maintained System Lumens	Maintained System Efficacy
1200	0.87	0.94	1963	39
1200	0.87	0.94	1963	43
1325 1400	0.91 0.93	0.93 0.93	2243 2422	52 56
3150	0.90	0.93	5273	58
1325 1400 1325 1400	0.91 0.93 0.91 0.93	0.92 0.92 0.75 0.75	2219 2396 1809 1953	60 65 67 72
1325 1400	0.91 0.93	0.92 0.92	3328 3594	64 69
1325 1400	0.91 0.93	0.92 0.92	4437 4791	63 68
3150 3150	0.90 0.90	0.87 0.90	4933 7655	67 72
1325 1400	0.91 0.93	0.90 0.90	2170 2344	66 71
1325 1400	0.91 0.93	0.92 0.92	3328 3594	71 76
1325 1400	0.91 0.93	0.90 0.90	4341 4687	70 76
3150 3150	0.90 0.90	0.90 0.88	5103 7484	72 74

NOTE: Maintained performance includes effect of lamp lumen depreciation.

TYPICAL PERFORMANCE VALUES

Ballast Types Lamp Types	Lamps per Ballast	Lamp Watts	System Input Watts	Lamp CRI
Old Standard Magnetic				
2 – F40T12	2	40	96	62
	2	40	96	73
	2	40	96	85
2 – F40T12/ES	2	34	82	62
	2	34	82	73
	2	34	82	85
2 – F40T10	2	40	101	82
Standard EE Magnetic				
2 – F40T12	2	40	88	62
	2	40	88	73
	2	40	88	85
2 – F40T12/ES	2	34	72	62
	2	34	72	73
	2	34	72	85
2 – F40T10	2	40	93	82
2 – F32T8	2	32	70	75
	2	32	70	85
Magnetic Heater Cutout				
2 – F40T12	2	40	80	62
	2	40	80	73
	2	40	80	85
Partial	2	40	69	62
Output	2	40	69	73
Ballast	2	40	69	85
2 – F40T12/ES	2	34	66	62
	2	34	66	73
	2	34	66	85

NOTE: Lamp lumen performance varies among manufacturers. Maintained performance includes effect of lamp lumen depreciation (@ 40% lamp life).

FOR 2-LAMP 4-FOOT SYSTEMS

Initial Lamp Lumens	Lamp Lumen Deprec.	Ballast Factor	Maintained System Lumens	Maintained System Efficacy	Maintained Relative Lumens
3050	0.87	0.94	4989	52	100%
3200	0.90	0.94	5414	56	109%
3300	0.90	0.94	5584	58	112%
2650	0.87	0.87	4012	49	80%
2800	0.90	0.87	4385	53	88%
2900	0.90	0.87	4541	55	91%
3700	0.89	0.92	6059	60	121%
3050	0.87	0.94	4989	57	100%
3200	0.90	0.94	5414	62	109%
3300	0.90	0.94	5584	63	112%
2650	0.87	0.87	4012	56	80%
2800	0.90	0.87	4385	61	88%
2900	0.90	0.87	4541	63	91%
3700	0.89	0.92	6059	65	121%
2850	0.91	0.94	4876	70	98%
3050	0.93	0.94	5333	76	107%
3050	0.87	0.95	5042	63	101%
3200	0.90	0.95	5472	68	110%
3300	0.90	0.95	5643	71	113%
3050	0.87	0.83	4405	64	88%
3200	0.90	0.83	4781	69	96%
3300	0.90	0.83	4930	71	99%
2650	0.87	0.88	4058	61	81%
2800	0.90	0.88	4435	67	89%
2900	0.90	0.88	4594	70	92%

TYPICAL PERFORMANCE VALUES

Ballast Types Lamp Types	Lamps per Ballast	Lamp Watts	System Input Watts	Lamp CRI
Magnetic Heater Cutout				
Partial	2	34	58	62
Output	2	34	58	73
Ballast	2	34	58	85
2 – F40T10	2	40	84	82
2 – F32T8	2	32	61	75
	2	32	61	85
Electronic Rapid Start				
2 – F40T12	2	40	72	62
	2	40	72	73
	2	40	72	85
2 – F40T12/ES	2	34	62	62
	2	34	62	73
	2	34	62	85
2 – F40T10	2	40	75	82
2 – F32T8	2	32	62	75
	2	32	62	85
Partial	2	32	54	75
Output	2	32	54	85
Extended	2	32	86	75
Output	2	32	86	85
Electronic Instant Start				
2 – F32T8	2	32	58	75
	2	32	58	85
Extended	2	32	76	75
Output	2	32	76	85

NOTE: Lamp lumen performance varies among manufacturers. Maintained performance includes effect of lamp lumen depreciation (@ 40% lamp life).

FOR 2-LAMP 4-FOOT SYSTEMS *(cont.)*

Initial Lamp Lumens	Lamp Lumen Deprec.	Ballast Factor	Maintained System Lumens	Maintained System Efficacy	Maintained Relative Lumens
2650	0.87	0.81	3735	64	75%
2800	0.90	0.81	4082	70	82%
2900	0.90	0.81	4228	73	85%
3700	0.89	0.92	6059	72	121%
2850	0.91	0.86	4461	73	89%
3050	0.93	0.86	4879	80	98%
3050	0.87	0.88	4670	65	94%
3200	0.90	0.88	5069	70	102%
3300	0.90	0.88	5227	73	105%
2650	0.87	0.88	4058	65	81%
2800	0.90	0.88	4435	72	89%
2900	0.90	0.88	4594	74	92%
3700	0.89	0.86	5664	76	114%
2850	0.91	0.88	4565	74	92%
3050	0.93	0.88	4992	81	100%
2850	0.91	0.75	3890	72	78%
3050	0.93	0.75	4255	79	85%
2850	0.91	1.28	6639	77	133%
3050	0.93	1.28	7261	84	146%
2850	0.91	0.88	4565	79	92%
3050	0.93	0.88	4992	86	100%
2850	0.91	1.15	5965	78	120%
3050	0.93	1.15	6524	86	131%

TYPICAL PERFORMANCE VALUES

Ballast Types Lamp Types	Lamps per Ballast	Lamp Watts	System Input Watts	Lamp CRI
Old Standard Magnetic				
3 – F40T12	1.5	40	148	62
	1.5	40	148	73
	1.5	40	148	85
3 – F40T12/ES	1.5	34	134	62
	1.5	34	134	73
	1.5	34	134	85
3 – F40T10	1.5	40	156	82
Standard EE Magnetic				
3 – F40T12	1.5	40	134	62
	1.5	40	134	73
	1.5	40	134	85
3 – F40T12/ES	1.5	34	112	62
	1.5	34	112	73
	1.5	34	112	85
3 – F40T10	1.5	40	142	82
3 – F32T8	1.5	32	106	75
	1.5	32	106	85
Magnetic Heater Cutout				
3 – F40T12	2/T	40	120	62
	2/T	40	120	73
	2/T	40	120	85
Partial	2/T	40	104	62
Output	2/T	40	104	73
Ballast	2/T	40	104	85

*NOTE: Lamp lumen performance varies among manufacturers. Maintained performance includes effect of lamp lumen depreciation (@ 40% lamp life).

FOR 3-LAMP 4-FOOT SYSTEMS

Initial Lamp Lumens	Lamp Lumen Deprec.	Ballast Factor	Maintained System Lumens	Maintained System Efficacy	Maintained Relative Lumens
3050	0.87	0.94	7483	51	100%
3200	0.90	0.94	8122	55	109%
3300	0.90	0.94	8375	57	112%
2650	0.87	0.87	6017	45	80%
2800	0.90	0.87	6577	49	88%
2900	0.90	0.87	6812	51	91%
3700	0.89	0.92	9080	58	121%
3050	0.87	0.94	7483	56	100%
3200	0.90	0.94	8122	61	109%
3300	0.90	0.94	8375	63	112%
2650	0.87	0.87	6017	54	80%
2800	0.90	0.87	6577	59	88%
2900	0.90	0.87	6812	61	91%
3700	0.89	0.92	9089	64	121%
2850	0.91	0.94	7314	69	98%
3050	0.93	0.94	7999	75	107%
3050	0.87	0.95	7562	63	101%
3200	0.90	0.95	8208	68	110%
3300	0.90	0.95	8465	71	113%
3050	0.87	0.83	6607	64	88%
3200	0.90	0.83	7171	69	96%
3300	0.90	0.83	7395	71	99%

NOTE: Ballast factors for electronic ballasts can vary in range of 0.41-1.30 among manufacturers.

TYPICAL PERFORMANCE VALUES

Ballast Types / Lamp Types	Lamps per Ballast	Lamp Watts	System Input Watts	Lamp CRI
Magnetic Heater Cutout				
3 – F40T12/ES	3	34	90	62
	3	34	90	73
(Partial Output)	3	34	90	85
3 – F40T10	2/T	40	142	82
3 – F32T8	2/T	32	122	75
	2/T	32	122	85
Electronic Rapid Start				
3 – F40T12	3	40	107	62
	3	40	107	73
	3	40	107	85
3 – F40T12/ES	3	34	92	62
	3	34	92	73
	3	34	92	85
3 – F40T10	3	40	116	82
3 – F32T8	3	32	90	75
	3	32	90	85
Partial	3	32	80	75
Output	3	32	80	85
Electronic Instant Start				
3 – F32T8	3	32	96	75
	3	32	96	85
	3	32	86	75
	3	32	86	85

NOTE: Lamp lumen performance varies among manufacturers. Maintained performance includes effect of lamp lumen depreciation (@ 40% lamp life).

FOR 3-LAMP 4-FOOT SYSTEMS *(cont.)*

Initial Lamp Lumens	Lamp Lumen Deprec.	Ballast Factor	Maintained System Lumens	Maintained System Efficacy	Maintained Relative Lumens
2650	0.87	0.83	5741	64	77%
2800	0.90	0.83	6275	70	84%
2900	0.90	0.83	6499	72	87%
3700	0.89	0.92	9089	64	121%
2850	0.91	0.86	6691	55	89%
3050	0.93	0.86	7318	60	98%
3050	0.87	0.88	7005	65	94%
3200	0.90	0.88	7603	71	102%
3300	0.90	0.88	7841	73	105%
2650	0.87	0.88	6087	66	81%
2800	0.90	0.88	6653	72	89%
2900	0.90	0.88	6890	75	92%
3700	0.89	0.92	9089	78	121%
2850	0.91	0.88	6847	76	92%
3050	0.93	0.88	7488	83	100%
2850	0.91	0.75	5835	73	78%
3050	0.93	0.75	6382	80	85%
2850	0.91	0.95	7391	77	99%
3050	0.93	0.95	8084	84	108%
2850	0.91	0.87	6769	79	90%
3050	0.93	0.87	7403	86	99%

NOTE: Ballast factors for electronic ballasts can vary in range of 0.41-1.30 among manufacturers.

TYPICAL PERFORMANCE VALUES

Ballast Types Lamp Types	Lamps per Ballast	Lamp Watts	System Input Watts	Lamp CRI
Old Standard Magnetic				
4 – F40T12	2	40	192	62
	2	40	192	73
	2	40	192	85
4 – F40T12/ES	2	34	164	62
	2	34	164	73
	2	34	164	85
Standard EE Magnetic				
4 – F40T12	2	40	176	62
	2	40	176	73
	2	40	176	85
4 – F40T12/ES	2	34	144	62
	2	34	144	73
	2	34	144	85
4 – F32T8	2	32	140	75
	2	32	140	85
Magnetic Heater Cutout				
4 – F40T12	2	40	160	62
	2	40	160	73
	2	40	160	85
Partial	2	40	138	62
Output	2	40	138	73
Ballast	2	40	138	85

NOTE: Lamp lumen performance varies among manufacturers. Maintained performance includes effect of lamp lumen depreciation (@ 40% lamp life).

FOR 4-LAMP 4-FOOT SYSTEMS

Initial Lamp Lumens	Lamp Lumen Deprec.	Ballast Factor	Maintained System Lumens	Maintained System Efficacy	Maintained Relative Lumens
3050	0.87	0.94	9977	52	100%
3200	0.90	0.94	10829	56	109%
3300	0.90	0.94	11167	58	112%
2650	0.87	0.87	8023	49	80%
2800	0.90	0.87	8770	53	88%
2900	0.90	0.87	9083	55	91%
3050	0.87	0.94	9977	57	100%
3200	0.90	0.94	10829	62	109%
3300	0.90	0.94	11167	63	112%
2650	0.87	0.87	8023	56	80%
2800	0.90	0.87	8770	61	88%
2900	0.90	0.87	9083	63	91%
2850	0.91	0.94	9752	70	98%
3050	0.93	0.94	10665	76	107%
3050	0.87	0.95	10083	63	101%
3200	0.90	0.95	10944	68	110%
3300	0.90	0.95	11286	71	113%
3050	0.87	0.83	8810	64	88%
3200	0.90	0.83	9562	69	96%
3300	0.90	0.83	9860	71	99%

NOTE: Ballast factors for electronic ballasts can vary in range of 0.41-1.30 among manufacturers.

TYPICAL PERFORMANCE VALUES

Ballast Types Lamp Types	Lamps per Ballast	Lamp Watts	System Input Watts	Lamp CRI
Magnetic Heater Cutout				
4 – F40T12/ES	2	34	132	62
	2	34	132	73
	2	34	132	85
Partial	2	34	116	62
Output	2	34	116	73
Ballast	2	34	116	85
4 – F32T8	2	32	122	75
	2	32	122	85
Electronic Rapid Start				
4 – F40T12	4	40	141	62
	4	40	141	73
	4	40	141	85
4 – F40T12/ES	4	34	117	62
	4	34	117	73
	4	34	117	85
4 – F32T8	4	32	116	75
	4	32	116	85
Partial	4	32	101	75
Output	4	32	101	85
Electronic Instant Start				
4 – F32T8	4	32	111	75
	4	32	111	85
Partial	4	32	101	75
Output	4	32	101	85

NOTE: Lamp lumen performance varies among manufacturers. Maintained performance includes effect of lamp lumen depreciation (@ 40% lamp life).

FOR 4-LAMP 4-FOOT SYSTEMS *(cont.)*

Initial Lamp Lumens	Lamp Lumen Deprec.	Ballast Factor	Maintained System Lumens	Maintained System Efficacy	Maintained Relative Lumens
2650	0.87	0.88	8115	61	81%
2800	0.90	0.88	8870	67	89%
2900	0.90	0.88	9187	70	92%
2650	0.87	0.81	7470	64	75%
2800	0.90	0.81	8165	70	82%
2900	0.90	0.81	8456	73	85%
2850	0.91	0.86	8922	73	89%
3050	0.93	0.86	9758	80	98%
3050	0.87	0.87	9234	65	93%
3200	0.90	0.87	10022	71	100%
3300	0.90	0.87	10336	73	104%
2650	0.87	0.83	7654	65	77%
2800	0.90	0.83	8366	72	84%
2900	0.90	0.83	8665	74	87%
2850	0.91	0.87	9025	78	90%
3050	0.93	0.87	9871	85	99%
2850	0.91	0.75	7781	77	78%
3050	0.93	0.75	8510	84	85%
2850	0.91	0.85	8818	79	88%
3050	0.93	0.85	9644	87	97%
2850	0.91	0.79	8195	81	82%
3050	0.93	0.79	8963	89	90%

NOTE: Ballast factors for electronic ballasts can vary in range of 0.41-1.30 among manufacturers.

TYPICAL PERFORMANCE VALUES

Ballast Types Lamp Types	Lamps per Ballast	Lamp Watts	System Input Watts	Lamp CRI
Old Standard Magnetic				
2 – F96T12	2	75	173	62
	2	75	173	73
	2	75	173	85
2 – F96T12/ES	2	60	138	62
	2	60	138	73
	2	60	138	85
2 – F96T12/HO	2	110	257	62
	2	110	257	73
	2	110	257	85
2 – F96T12/HO/ES	2	95	227	62
	2	95	227	73
	2	95	227	85
2 – F96T12/VHO	2	215	450	62
2 – F96T12/VHO/ES	2	185	390	62
Standard EE Magnetic				
2 – F96T12	2	75	158	62
	2	75	158	73
	2	75	158	85
2 – F96T12/ES	2	60	128	62
	2	60	128	73
	2	60	128	85
2 – F96T12/HO	2	110	237	62
	2	110	237	73
	2	110	237	85
2 – F96T12/HO/ES	2	95	197	62
	2	95	197	73
	2	95	197	85

NOTE: Lamp lumen performance varies among manufacturers. Maintained performance includes effect of lamp lumen depreciation (@ 40% rated life).

FOR 2-LAMP 8-FOOT SYSTEMS

Initial Lamp Lumens	Lamp Lumen Deprec.	Ballast Factor	Maintained System Lumens	Maintained System Efficacy	Maintained Relative Lumens
6100	0.88	0.94	10092	58	100%
6425	0.94	0.94	11354	66	113%
6600	0.94	0.94	11664	67	116%
5500	0.88	0.87	8422	61	83%
5750	0.94	0.87	9405	68	93%
5900	0.94	0.87	9650	70	96%
8900	0.87	0.94	14557	57	144%
9200	0.90	0.94	15566	61	154%
9400	0.90	0.94	15905	62	158%
8000	0.87	0.87	12110	53	120%
8350	0.90	0.87	13076	58	130%
8600	0.90	0.87	13468	59	133%
13500	0.75	0.94	19035	42	189%
12500	0.75	0.87	16313	42	162%
6100	0.88	0.95	10199	65	100%
6425	0.94	0.95	11475	73	114%
6600	0.94	0.95	11788	75	117%
5500	0.88	0.90	8712	68	86%
5750	0.94	0.90	9729	76	96%
5900	0.94	0.90	9983	78	99%
8900	0.87	0.95	14712	62	146%
9200	0.90	0.95	15732	66	156%
9400	0.90	0.95	16074	68	159%
8000	0.87	0.88	12250	62	121%
8350	0.90	0.88	13226	67	131%
8600	0.90	0.88	13622	69	135%

TYPICAL PERFORMANCE VALUES

Ballast Types Lamp Types	Lamps per Ballast	Lamp Watts	System Input Watts	Lamp CRI
Standard EE Magnetic				
2 – F96T12/VHO	2	215	440	62
2 – F96T12/VHO/ES	2	185	380	62
Magnetic Heater Cutout				
2 – F96T12/HO	2	110	210	62
	2	110	210	73
	2	110	210	85
2 – F96T12/HO/ES	2	95	177	62
	2	95	177	73
	2	95	177	85
Electronic				
2 – F96T12	2	75	136	62
	2	75	136	73
	2	75	136	85
2 – F96T12/ES	2	60	110	62
	2	60	110	73
	2	60	110	85
2 – F96T12/HO	2	110	209	62
	2	110	209	73
	2	110	209	85
2 – F96T12/HO/ES	2	95	174	62
	2	95	174	73
	2	95	174	85
2 – F96T8	2	59	106	75
	2	59	106	84
Partial	2	59	99	75
Output	2	59	99	84
2 – F96T8/HO	2	86	160	85

NOTE: Lamp lumen performance varies among manufacturers. Maintained performance includes effect of lamp lumen depreciation (@ 40% rated life).

FOR 2-LAMP 8-FOOT SYSTEMS *(cont.)*

Initial Lamp Lumens	Lamp Lumen Deprec.	Ballast Factor	Maintained System Lumens	Maintained System Efficacy	Maintained Relative Lumens
13500	0.75	0.95	19238	44	191%
12500	0.75	0.88	16500	43	163%
8900	0.87	0.89	13783	66	137%
9200	0.90	0.89	14738	70	146%
9400	0.90	0.89	15059	72	149%
8000	0.87	0.86	11971	68	119%
8350	0.90	0.86	12926	73	128%
8600	0.90	0.86	13313	75	132%
6100	0.88	0.89	9555	70	100%
6425	0.94	0.89	10750	79	107%
6600	0.94	0.89	11043	81	109%
5500	0.88	0.88	8518	77	84%
5750	0.94	0.88	9513	86	94%
5900	0.94	0.88	9761	89	97%
8900	0.87	0.90	13937	67	138%
9200	0.90	0.90	14904	71	148%
9400	0.90	0.90	15228	73	151%
8000	0.87	0.88	12250	70	121%
8350	0.90	0.88	13226	76	131%
8600	0.90	0.88	13622	78	135%
5800	0.91	0.85	8973	85	89%
5950	0.91	0.85	9205	87	91%
5800	0.91	0.78	8234	83	82%
5950	0.91	0.78	8447	85	84%
8200	0.90	0.88	12989	81	129%

PREHEAT CIRCUITS

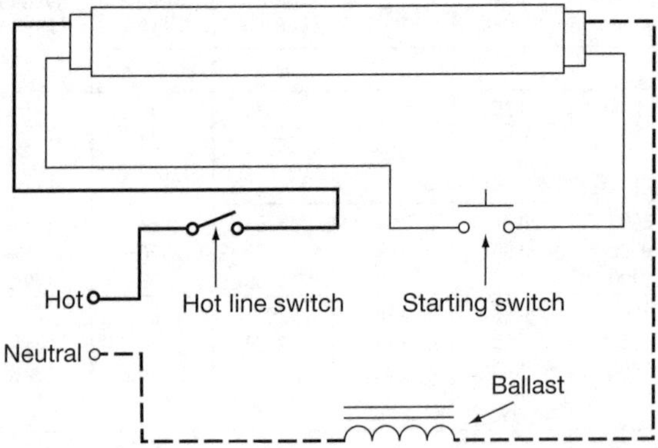

BALLAST

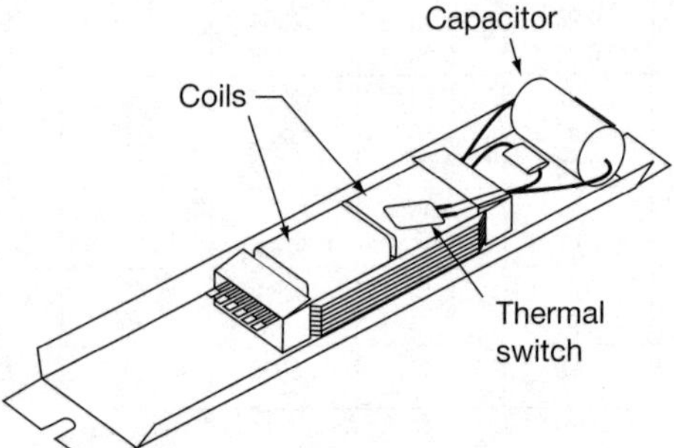

RAPID-START CIRCUITS

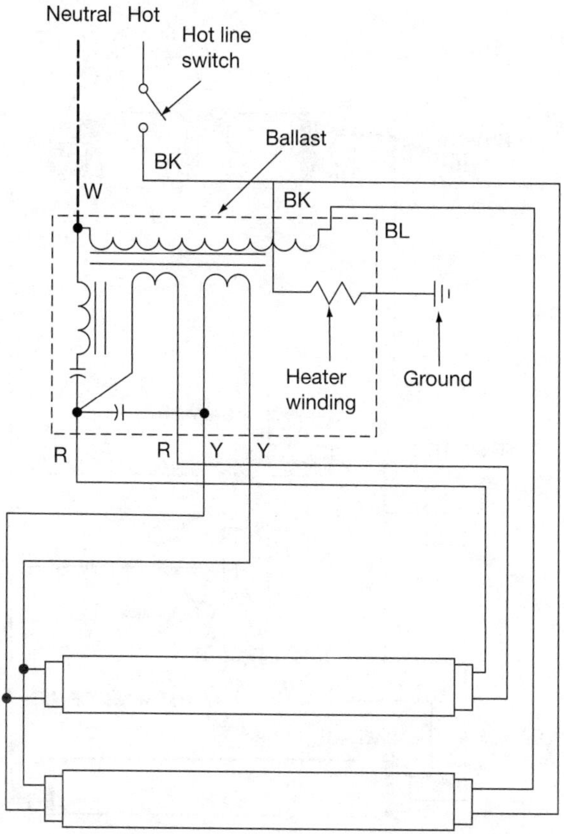

INSTANT-START CIRCUITS

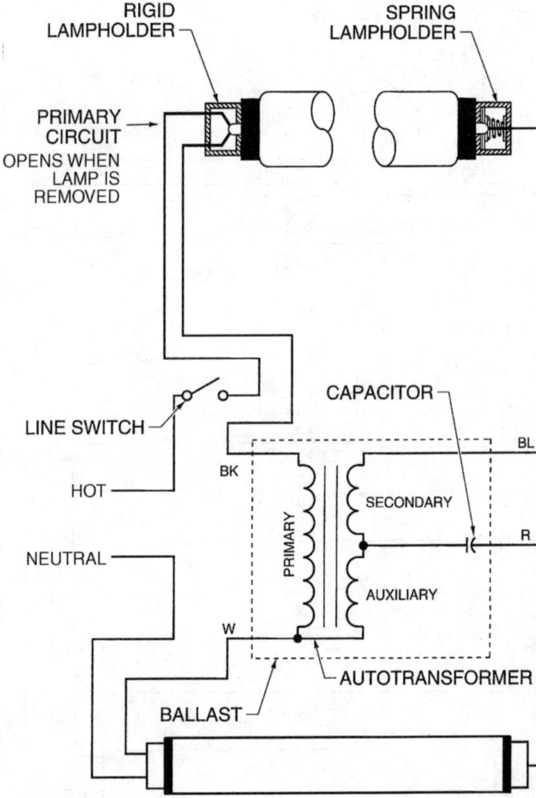

CHARACTERISTICS OF COMMON BALLASTS

	Input (watts)	Ballast Factor	Efficacy (lum./watt)	Power Factor	Current THD
Ballast for 2-F32T8 fluorescent					
Energy-efficient magnetic	70	0.94	78	≥ 0.9	≤ 20%
Hybrid	61	0.86	82	≥ 0.9	≤ 20%
Electronic	62	0.88	82	≥ 0.9	≤ 20%
Electronic, reduced light output	51	0.71	81	≥ 0.9	≤ 20%
Instant-start electronic	63	0.95	87	≥ 0.9	≤ 20%
Ballast for 2-F34T12 fluorescent					
Energy-efficient magnetic	72	0.87	68	≥ 0.9	≤ 20%
Hybrid	66	0.88	75	≥ 0.9	≤ 20%
Hybrid, reduced light output	58	0.81	78	≥ 0.9	≤ 20%
Electronic	62	0.88	79	≥ 0.9	≤ 20%
Electronic, reduced light output	52	0.73	79	≥ 0.9	≤ 20%

U.S. AND CANADIAN STANDARDS FOR BALLAST EFFICACY FACTOR

Lamp Type	Ballast Input Voltage (V)	Nominal Active Power (W)	BEF (minimum) U.S.	BEF (minimum) Canada
One F40T12	120	40	1.805	1.805
	277	40	1.805	1.805
	347	40	N/A	1.750
Two F40T12	120	80	1.060	1.060
	277	80	1.050	1.050
	347	80	N/A	1.020
Two F96T12	120	150	0.570	0.570
	277	150	0.570	0.570
	347	150	N/A	0.560
Two F96T12/HO	120	220	0.390	0.390
	277	220	0.390	0.390
	347	220	N/A	0.380
Two F32T8	120	64	N/A	1.250
	277	64	N/A	1.230
	347	64	N/A	1.200

TYPICAL MAGNETIC BALLAST WATTS

Lamp Type	Lamp Watts	Standard Ballast Watts	Standard System Watts	Energy Saving Ballast Watts	Energy Saving System Watts
1 Lamp/ballast					
F4T5	4	5	9	–	–
F20T12	20	5	25	–	–
2 Lamps/ballast					
F30T12	30	21	81	–	–
F40T12	40	16	96	6	86
F96T12	75	23	173	8	158
F96T12/HO	110	37	257	17	237

APPROXIMATE ANSI THERMAL CORRECTION FACTORS: WATTAGE

Type	Parabolic	Lens	Air Return	Strip
40W/T12/Magnetic	0.92	0.91	0.99	1.00
34W/T12/Magnetic	0.98	0.95	1.00	1.00
40W/T12/Electronic	0.94	0.92	0.99	1.00
34W/T12/Electronic	1.00	0.97	1.02	1.00
T8/Magnetic	0.95	0.92	0.98	1.00
T8/Electronic	0.94	0.90	0.98	1.00

Note: Luminaires (except strip fixtures) are assumed to be recessed in grid ceiling. Thermally corrected wattage=ANSI wattage x correction factor

APPROXIMATE ANSI THERMAL CORRECTION FACTORS: LUMENS

Type	Parabolic	Lens	Air Return	Strip
40W/T12/Magnetic	0.96	0.96	1.11	1.00
34W/T12/Magnetic	0.98	0.95	1.09	1.00
40W/T12/Electronic	0.97	0.97	1.09	1.00
34W/T12/Electronic	0.99	0.97	1.07	1.00
T8/Magnetic	0.98	0.96	1.07	1.00
T8/Electronic	0.98	0.95	1.08	1.00

Note: Luminaires (except strip fixtures) are assumed to be recessed in grid ceiling. Thermally corrected lumens = ANSI lumens x correction factor

FLUORESCENT LAMP BASES

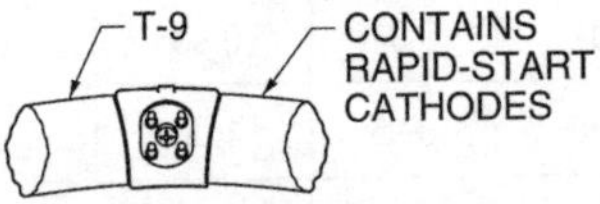

FOUR-PIN

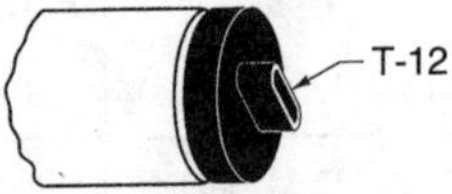

RECESSED DOUBLE CONTACT

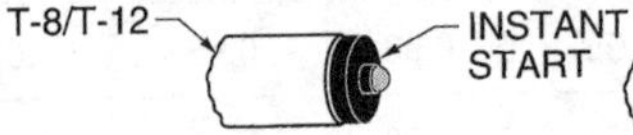

SINGLE-PIN

MINIATURE BI-PIN

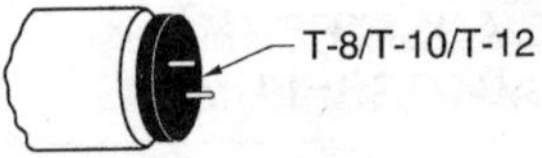

MEDIUM BI-PIN

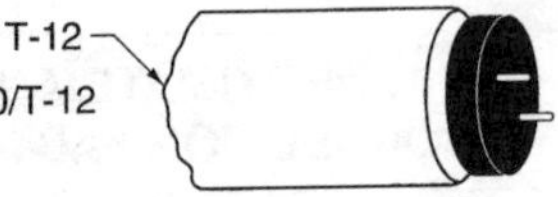

MOGUL BI-PIN

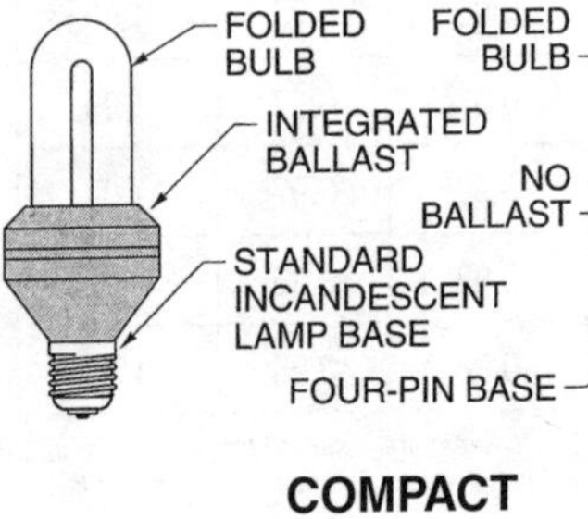

COMPACT

TYPICAL FLUORESCENT BASES WITH ANSI DESIGNATIONS

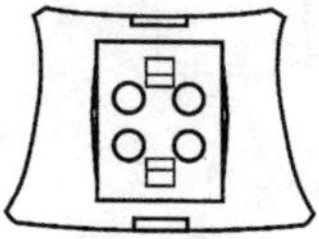

4-Pin
G10q

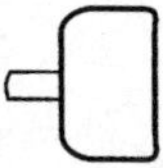

Single pin
Fa8

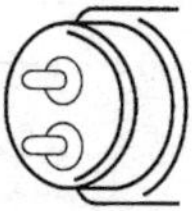

Medium bi-pin
G13

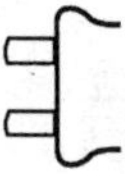

Miniature bi-pin
G5

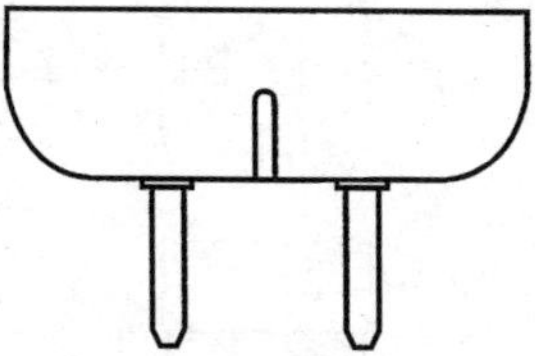

Medium bi-pin
G20

TYPICAL CFL BASES WITH ANSI DESIGNATIONS

Medium screw
E26

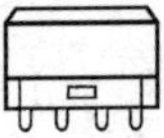

2G11-4

G23

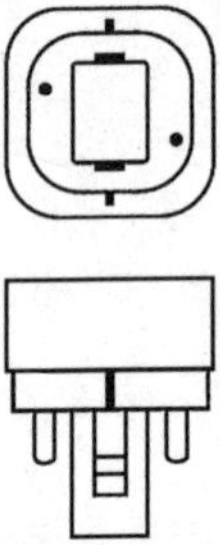

G23-2

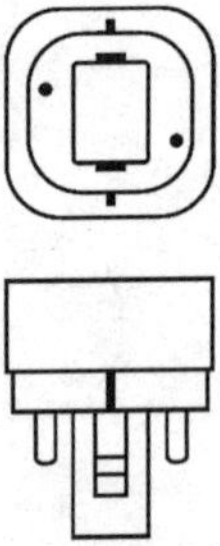

G24d-1

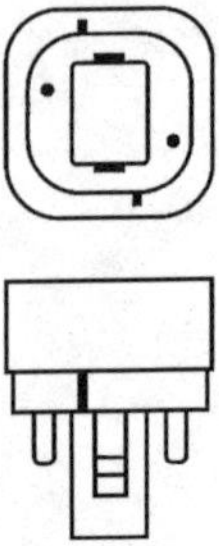

G24-2

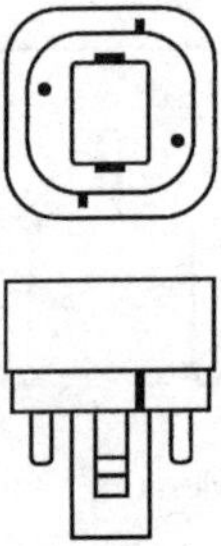

G24d-3

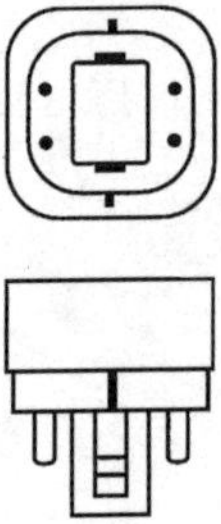

G24q-1

TYPICAL CFL BASES WITH ANSI DESIGNATIONS *(cont.)*

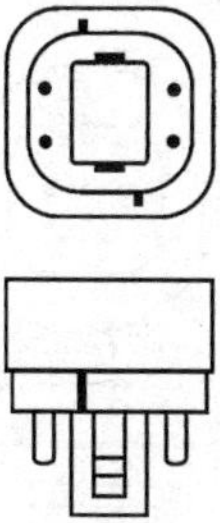

G24q-2

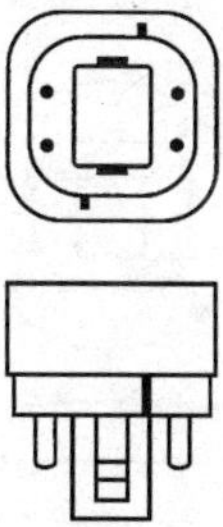

G24q-3

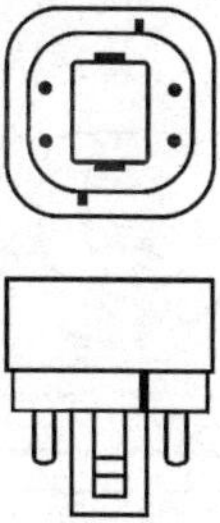

G24q-3D

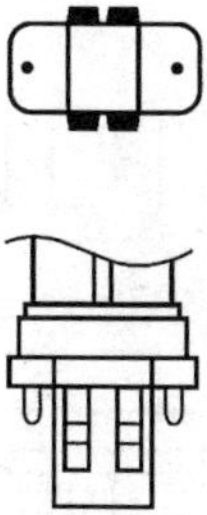

GX23

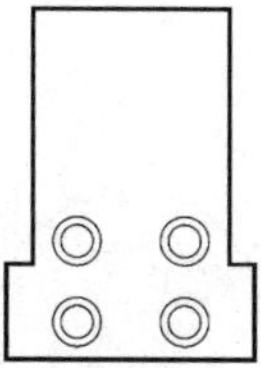

GR10q-4

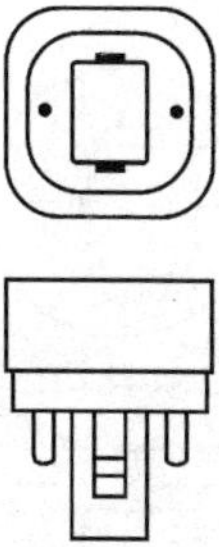

GX23-2

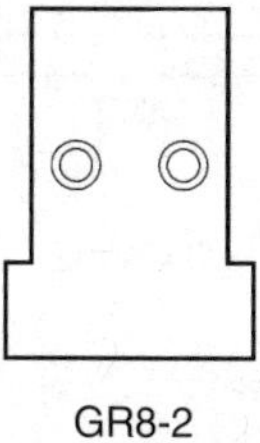

GR8-2

TYPICAL CFL BASES WITH ANSI DESIGNATIONS *(cont.)*

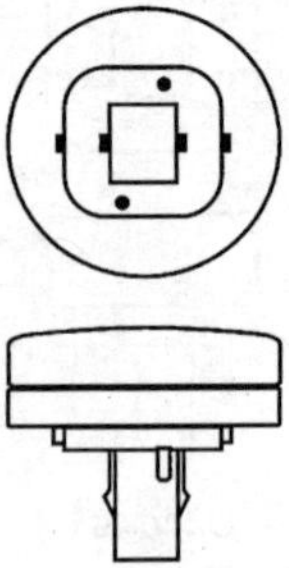

G24d-1

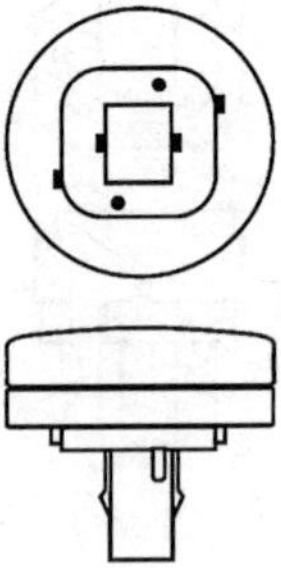

GX24d-2

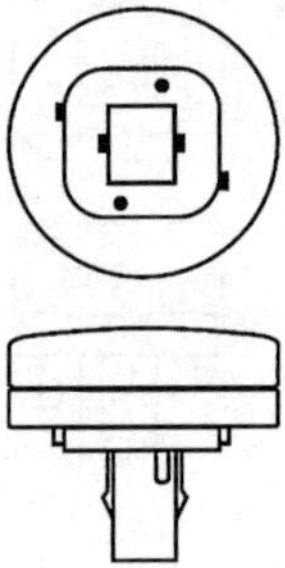

GX24d-3

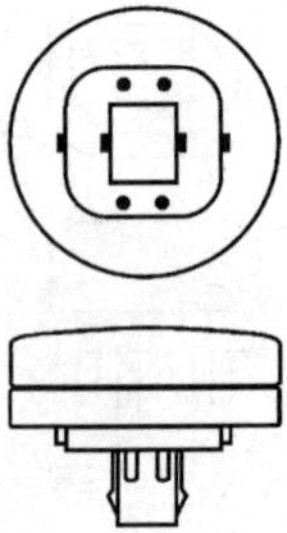

GX24q-1

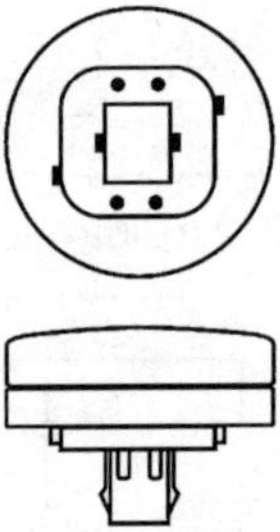

GX24q-2

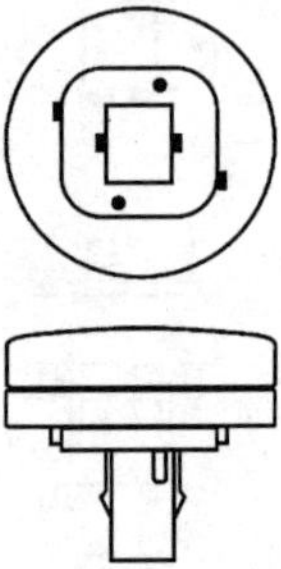

GX24q-3

FLUORESCENT LAMP TROUBLESHOOTING GUIDE

Problem	Circuit Type	Possible Cause	Corrective Action
Lamp blinks & has shimmering effect during lighting period	All types	Depletion of emission material on electrodes.	Replace with new lamp.
New lamp blinks	All types	Loose lamp connection.	Reseat lamp in socket securely. Ensure lamp holders are rigidly mounted and properly spaced.
		Low voltage applied to circuit.	Ensure that voltage is within ±7% of rated voltage.
		Cold area or draft hitting lamp.	Protect lamp with enclosure. Use a low temperature-rated ballast.
	Preheat	Defective or old starter.	Replace starter. Lamp life is reduced when starter is not replaced.
Lamp does not light or is slow starting	All types	Lamp failure.	Replace lamp. Test faulty lamp in another fixture. Check circuit fuse or CB. Check voltage at fixture.
		Loss of power or low voltage to the fixture.	Troubleshoot fluorescent fixture when voltage is present at the fixture. Replace broken or cracked holder. Check for poor wire connection.
	Preheat Rapid-start	Normal end of starter life Failed capacitor in ballast.	Replace starter. Replace ballast.
Lamp does not light	All types	Lamp not seated in holder.	Seat lamp properly in holder. In rapid-start circuits, the holder includes a switch that removes power when the lamp is removed due to high voltage present.
Bulb ends remain lighted after switch is turned OFF	Preheat	Starter contacts stuck together.	Replace starter. Lamp life is reduced when starter is not replaced.

FLUORESCENT LAMP TROUBLESHOOTING GUIDE *(cont.)*

Problem	Circuit Type	Possible Cause	Corrective Action
Short lamp life	All types	Frequent turning ON and OFF of lamps.	Normal operation is based on one start per three-hour period of operation time. Short lamp life must be expected when frequent starting cannot be avoided.
		Supply voltage excessive or low.	Check the supply voltage against the ballast rating. Short lamp life must be expected when supply voltage is not within ±7% of lamp rating.
		Low ambient temperature. Low temperature causes a slow start.	Protect lamp with enclosure. Use a low temperature-rated ballast.
	Instant-start	One lamp burned out and other burning dimly due to series-start ballast circuit.	Replace burned-out lamp. Ballast is damaged when lamp is not replaced.
		Wrong lamp type. May be using rapid-start or pre-heat lamp instead of instant-start.	Replace lamp with correct type.
Light output decreases	All types	Light output decreases over first 100 hours of operation.	Rated light output is based on output after 100 hours of lamp operation. Before 100 hours of operation, light output may be as much as 10% higher than normal.
		Low circuit voltage.	Check the supply voltage against the ballast rating. Short lamp life and low light output must be expected when supply voltage is not within ±7% of lamp rating.
		Dirt build-up on lamp and fixture.	Clean bulb and fixture.

CHAPTER 4
INCANDESCENTS

INCANDESCENT

With efficacies ranging from 6 to 24 lumens per watt, incandescent lamps are the least energy-efficient light source and have a relatively short life (750-2500 hours).

TUNGSTEN

The tungsten halogen lamp is another type of incandescent. In halogens, a small quartz capsule contains the filament and halogen gas. The small size allows the filament to operate at a higher temperature, producing light at a higher efficacy than incandescents. The halogen gas combines with evaporated tungsten and is redeposited on the filament thus extending the life of the filament while keeping the bulb wall from blackening.

Because the filament is small, this source is used where a focused beam is desired. Compact halogens are popular in retail store displays and accent lighting. Tungsten-halogen lamps generally produce a whiter light than incandescents, are more efficient, last longer, and have improved lamp lumen depreciation.

COMPACT FLUORESCENTS

Note that pages 4-18 through 4-23 compare incandescent light sources with compact fluorescent light sources.

COMMON INCANDESCENT LAMPS			
Type	Watts	Size (in)	Volts
A-19	40-100	2⅜	120-130
F-15	25-60	1⅞	120
PS-35	300-500	4⅜	120
PAR-38	115-150	4¾	120-130
T-19	40-100	2⅜	120

INCANDESCENT LAMP DATA

Description	Base	Watts	Lumens	Life (hrs)
A-19	Med	100	1750	750
A-21	Med	150	2880	750
A-23	Med	200	4010	750
PS-35	Mogul	300	5820	1000
PS-35	Mogul	500	10850	1000

TYPE A INCANDESCENT BULB RATINGS

Lamp Rating (watts)	Approximate Lumens
15	125
25	250
40	480
60	740
75	1210
100	1750
150	2850
200	3940

EFFICACY OF COMMON INCANDESCENT LAMPS

Watts	Description	Volts	Life (hrs)	Lumens	LM/Watt
25	General Service	120	2500	235	9.4
40	General Service	120	1000	480	12.0
40	Traffic Signal	120	2000	380	9.5
60	General Service	120	1000	870	14.5
75	General Service	120	750	1190	15.9
100	General Service	120	750	1750	17.5
100	General Service	120	2500	1440	14.4
100	General Service	120	4000	1290	12.9
100	Train	34	1000	2160	21.6
100	General Service	230	1000	1280	12.8
150	General Service	120	750	2850	19.0
300	General Service	120	750	6360	21.2
500	General Service	120	1000	10850	21.7
1000	General Service	120	1000	23740	23.7

Note that the efficacy increases as wattage increases. Note also that long life lamps have lower efficacy than standard life rated lamps, and that low-voltage lamps are more efficient than high-voltage lamps. Lamps designed to operate at high voltages have thinner filaments which are less efficient due to higher gas losses.

INCANDESCENT LAMPS OPERATED BELOW RATED VOLTAGE

Input to Lamp (volts)	Light Output (percent)	Watts (percent)	Color Change (kelvin)	Life (percent)	Relative Efficacy
120	100	100	0	100	1.00
110	75	88	-100	300	0.85
100	55	76	-200	1000	0.72
90	38	64	-300	4000	0.59

U.S. EPACT INCANDESCENT REFLECTOR LAMP STANDARD LEVELS

National Lamp Wattage	Minimum Average Lamp Efficacy (lumens/watts)
40-50	10.5
51-66	11.0
67-85	12.5
86-115	14.0
116-155	14.5
156-205	15.0

TYPE T TUNGSTEN-HALOGEN BULB RATINGS	
Wattage Rating of Bulb	**Light Output of Bulb**
5 watts	55 lumens
10 watts	120 lumens
20 watts	350 lumens
35 watts	650 lumens
50 watts	950 lumens
75 watts	1350 lumens
100 watts	2300 lumens
150 watts	3200 lumens

INCANDESCENT LAMP CHARACTERISTICS

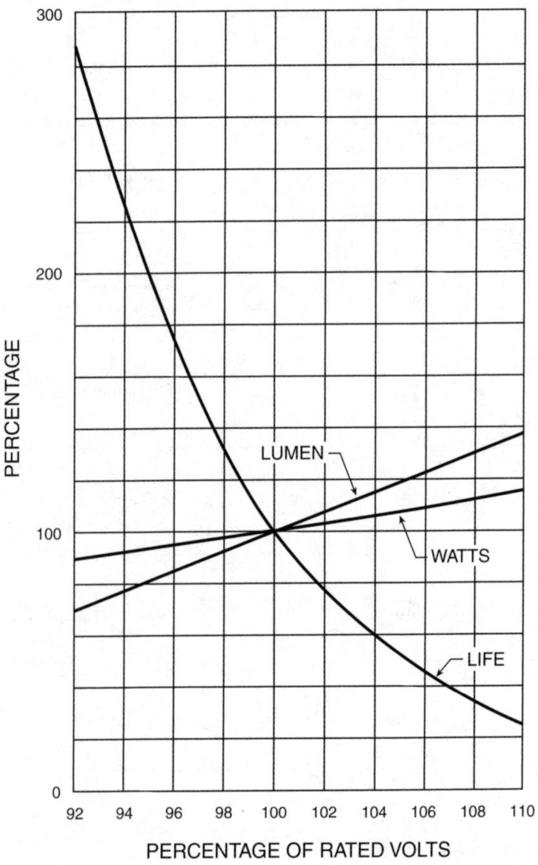

REDUCTIONS IN LIGHT OUTPUT AND POWER CONSUMPTION OF A TYPICAL INCANDESCENT LAMP OVER LIFE

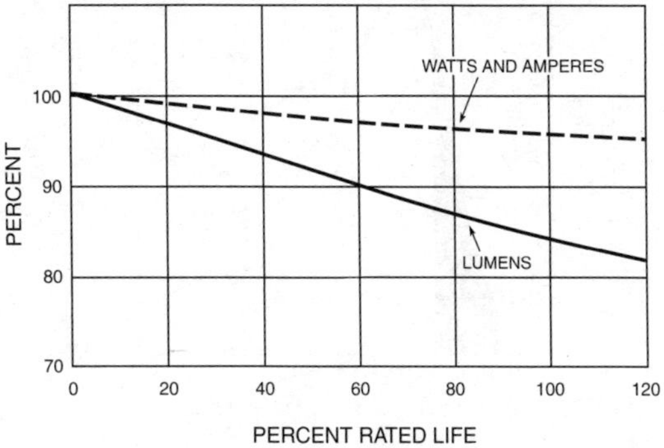

MORTALITY CURVE FOR TYPICAL INCANDESCENT LAMPS

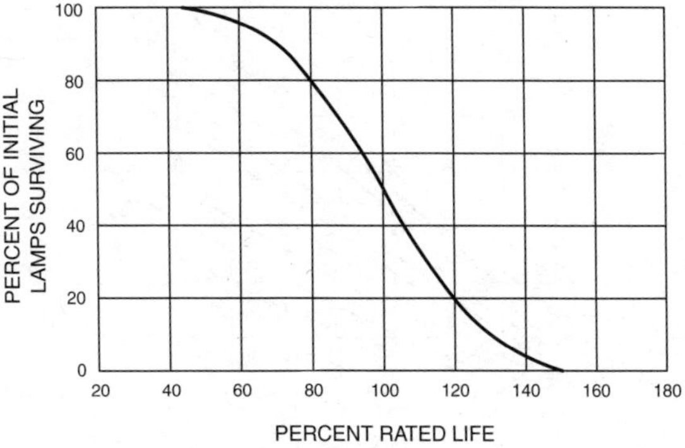

SPECTRAL ENERGY DISTRIBUTION FOR A TYPICAL INCANDESCENT LAMP OPERATING AT A FILAMENT TEMPERATURE OF 3000° K

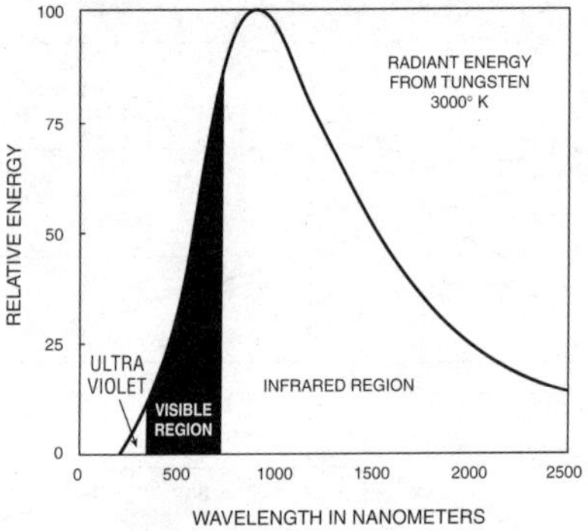

RELATIVE ENERGY RADIATED AS LIGHT FROM INCANDESCENT LAMPS AT VARIOUS TEMPERATURES

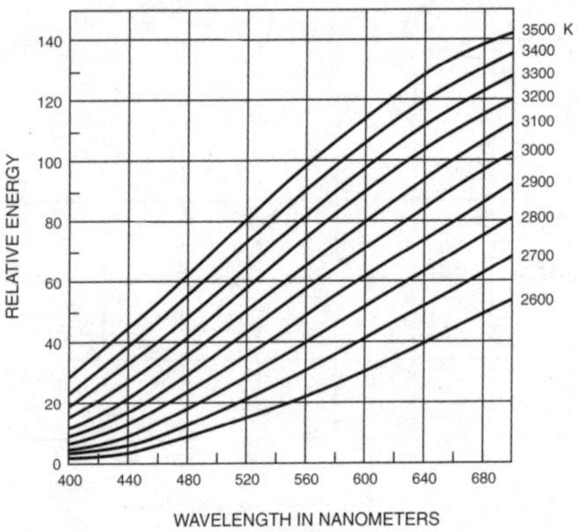

INCANDESCENT BASES WITH ANSI DESIGNATIONS

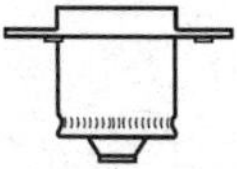

Single contact prefocus

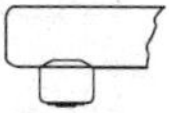

S-14S Philinea

Disc base

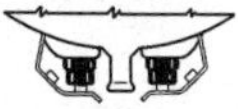

Screw terminal

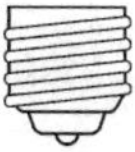

Intermediate E17

Candelabra screw E12

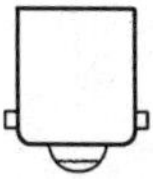

Candelabra SC bay B15

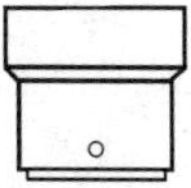

2 Lug sleeve B22d

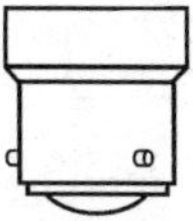

3 Lug sleeve B22-3

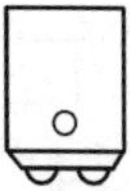

Candelabra DC bay

3 Contact medium E26d

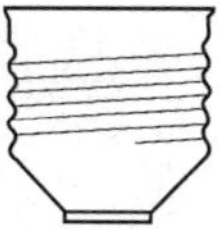

Medium screw E26

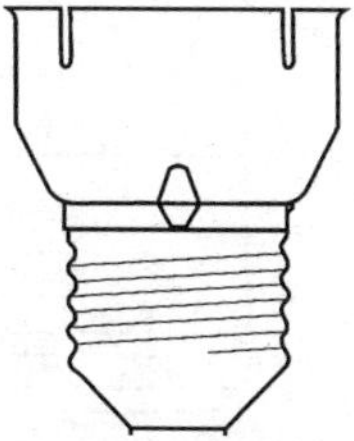

Medium skirt E26/50x39

INCANDESCENT BASES WITH ANSI DESIGNATIONS *(cont.)*

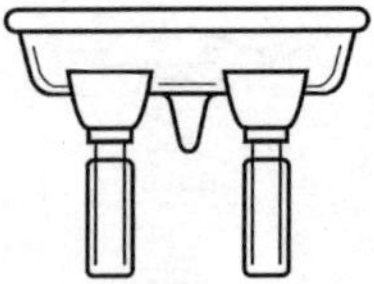

Medium bipost
G22

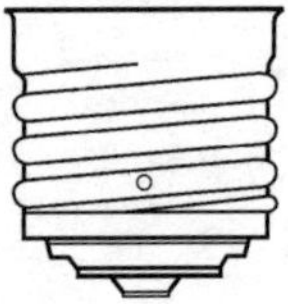

3 Contact Mogul
E39D

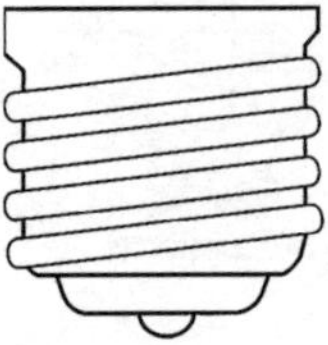

Mogul screw
E39

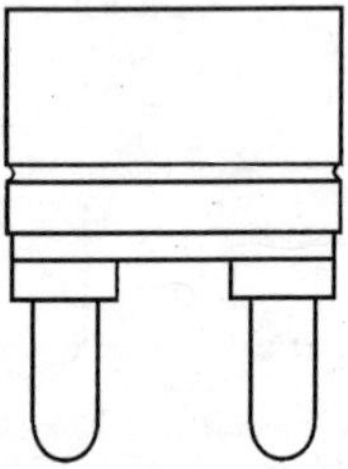

Mogul bipost
G38

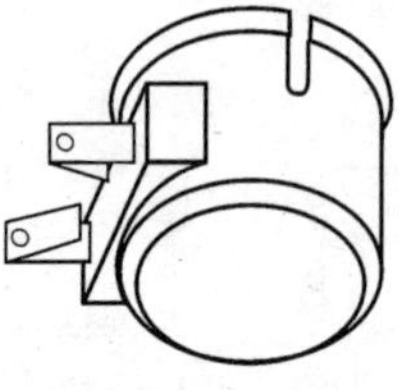

Medium
side prong

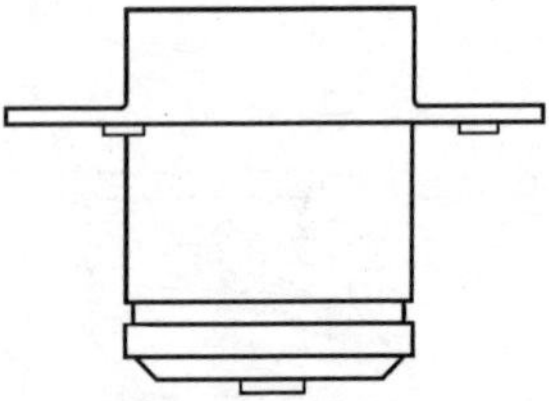

Medium prefocus
P28S

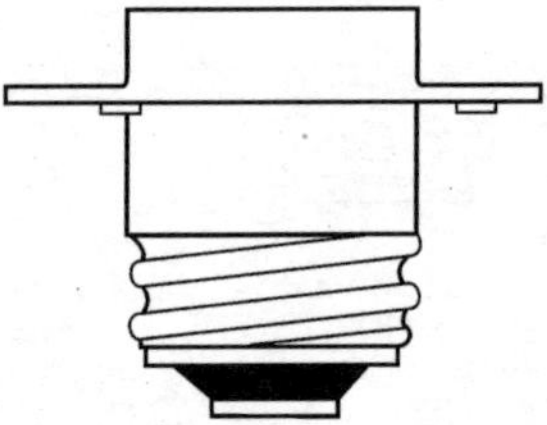

Mogul
P40S

INCANDESCENT BASES WITH ANSI DESIGNATIONS *(cont.)*

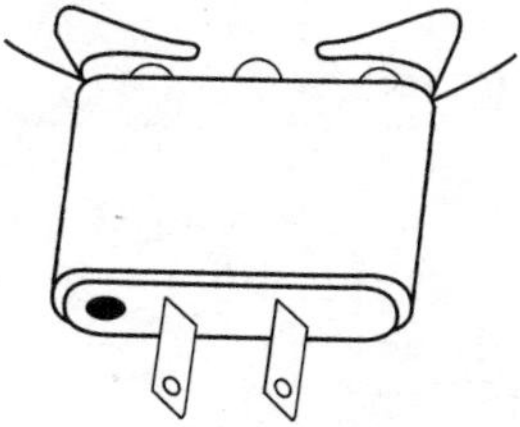

Extended mogul
end prong
GX16

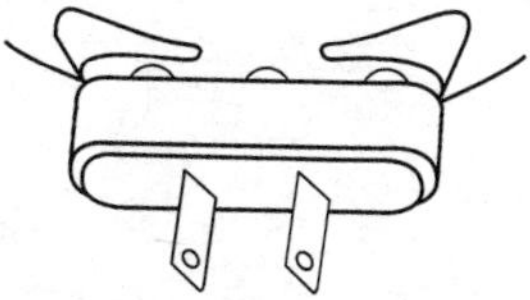

Mogul
end prong

HALOGEN BASES WITH ANSI DESIGNATIONS

2-Pin
prefocus

Can DC
bay

Miniature
screw
E10

Miniature
candelabra
E11

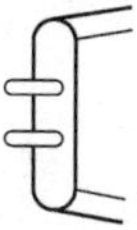

2-Pin
round
GX5.3

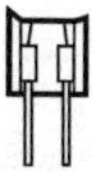

2-Pin
GY6.35

2-Pin
G-4

2-Pin
GU-4

2-Pin
GU-5.3

HALOGEN BASES WITH ANSI DESIGNATIONS *(cont.)*

4" Leads

6" Flex leads

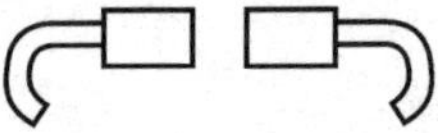

1" Ribbon leads

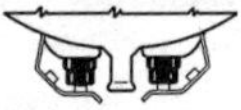

Screw
terminal

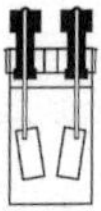

Turn & lock
GU-7

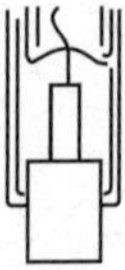

Recessed
single
contact
R7S

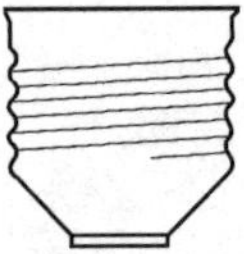

Medium
screw
E26

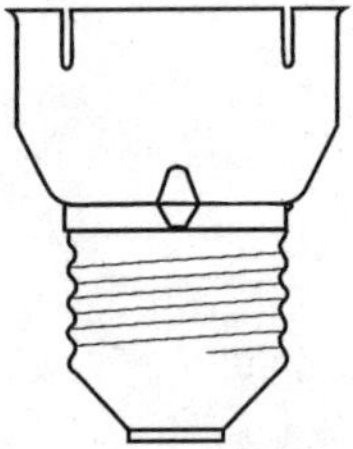

Medium skirted
E26/50x39

HALOGEN BASES WITH ANSI DESIGNATIONS *(cont.)*

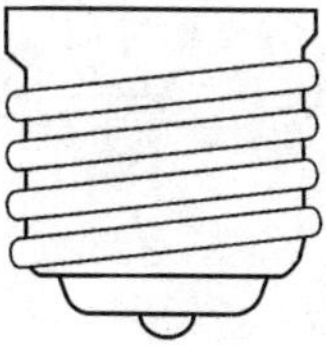

Mogul screw
E39

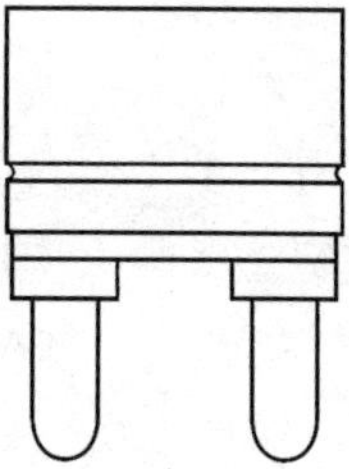

Mogul bipost
G38

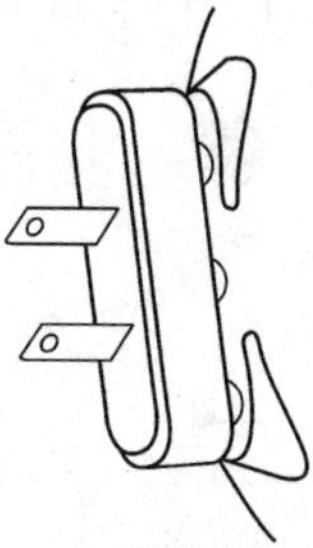

Mogul end prong
GX16d

Extended mogul
end prong
GX16d

TYPICAL BULB SHAPES AND THEIR ANSI DESIGNATIONS

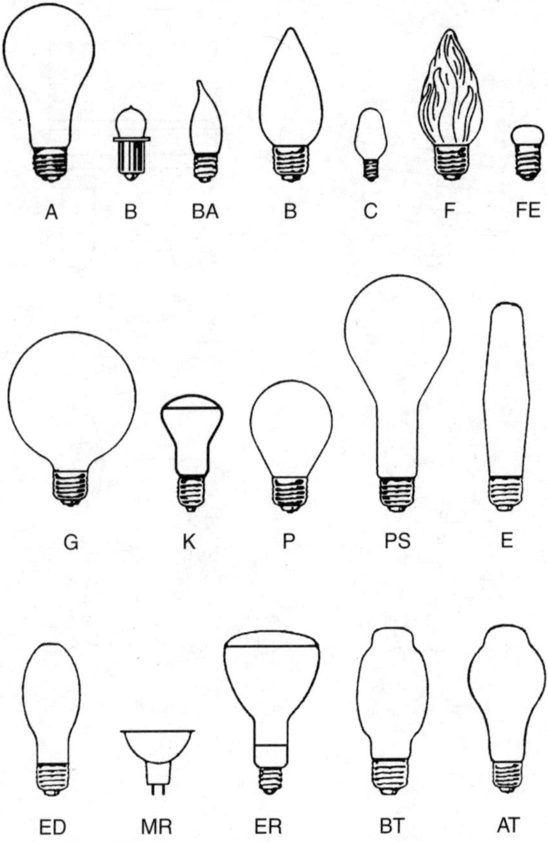

TYPICAL BULB SHAPES AND THEIR ANSI DESIGNATIONS *(cont.)*

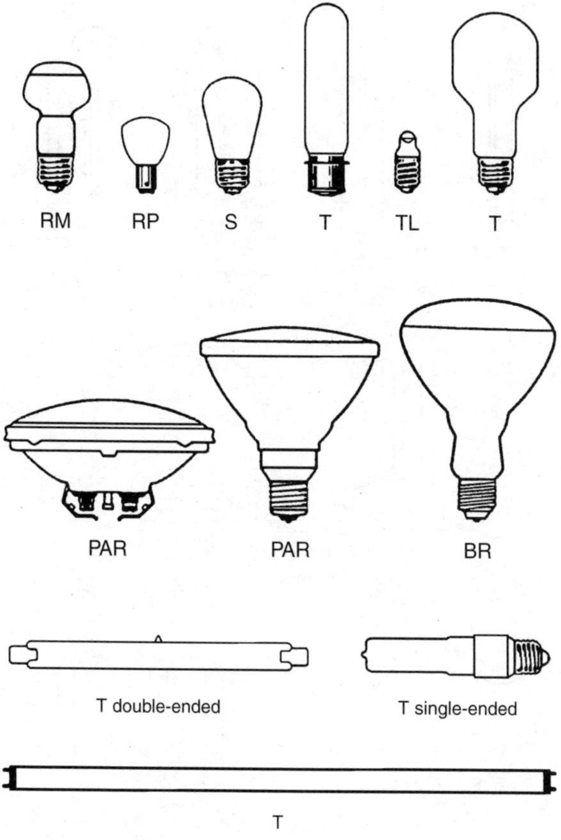

TYPICAL BULB SHAPES AND THEIR ANSI DESIGNATIONS *(cont.)*

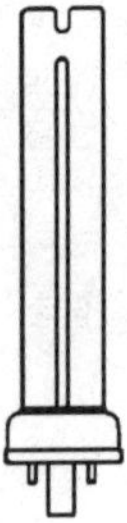

2-pin

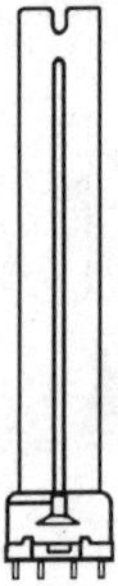

4-pin

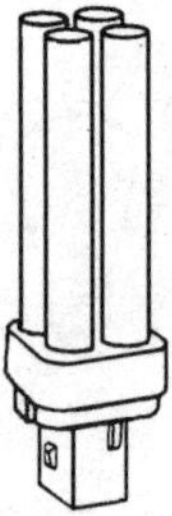

Double

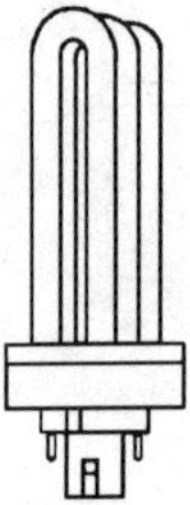

Triple

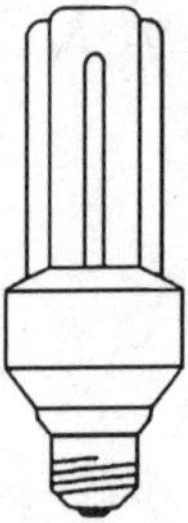

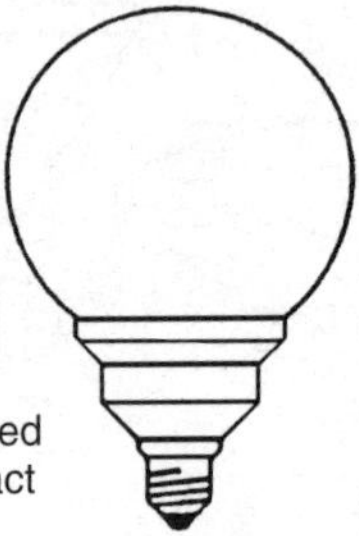

Ballasted compact

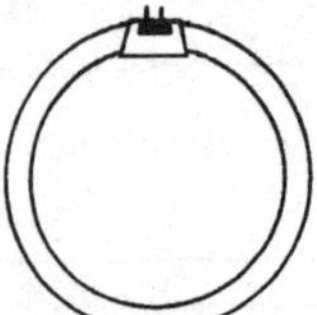

T/C

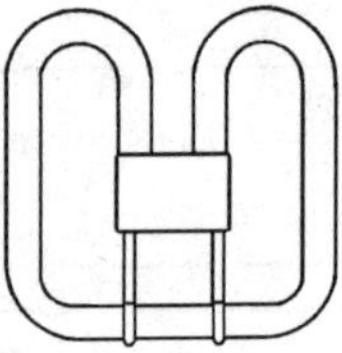

2D

TYPICAL BULB SHAPES AND THEIR ANSI DESIGNATIONS *(cont.)*

Flat

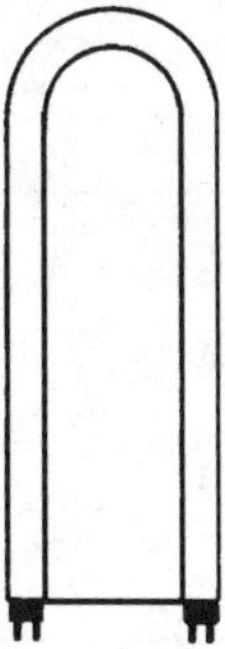

TU

ANSI BULB SHAPE DESCRIPTIONS

A – Arbitrary spherical shape tapered
AT – Arbitrary tubular
B – Bulged or bullet shape; blunt tip
BA – Bulged with angular (bent) tip
BD – Bulged with dimple in crown
BR – Bulged reflector
BT – Bulged tubular
C – Conical
CA – Candle shape with bent tip
CC – Two conical shapes blended
E – Elliptical
ED – Elliptical with dimple in crown
ER – Elliptical reflector
F – Flame shape, decorative
FE – Flat elliptical
G – Globe shape
GT – Globe/tubular combination
K – Similar to M with conical transition
M – Mushroom shape/rounded transitions
MR – Multifaceted reflector
P – Pear shape
PS – Pear shape with straight neck
PAR – Parabolic aluminized reflector
R – Reflector
RB – Bulged reflector
RD – Reflector with dimple in crown
REC – PAR type with rectangular face
RM – Reflector, mushroom shape
RP – Reflector, pear shape
S – Straight-sided shape
ST – Straight-tipped shape
T – Tubular shape
TL – Tubular shape with lens in crown
T/C – Tubular circular
TU – Tubular U-shape
2D – 2-dimensional

TYPICAL PERFORMANCE VALUES FOR COMPACT SOURCES

(non-directional sources)

Lamp Types	Lamp Watts	System Watts	Initial Lamp CRI	Maintained Lamp Lumens	Maintained System Lumens	System Efficacy	Rated Life
Incandescents (Reference)							
A19	25	25	100	215	215	9	1000
A19	40	40	100	495	495	12	1000
A19	60	60	100	860	860	14	1000
A19	75	75	100	1180	1180	16	750
A19	100	100	100	1720	1720	17	750
A23	200	200	100	4010	4010	20	750
TB19 (halogen)	50	50	100	830	830	17	2000
TB19 (halogen)	90	90	100	1680	1680	19	2000
TB19 (halogen/film)	60	60	100	1400	1400	23	2000
Integral Units/Electronic							
Enclosed/Non Vent	–	16	82	800	650	41	10000
Enclosed/Vented	–	18	82	1100	950	53	10000
Open Quad Tube	–	15	82	900	765	51	10000
Open Quad Tube	–	20	82	1200	1020	51	10000

NOTES: Lamp lumen performance varies among manufacturers.
Maintained performance includes effect of lamp lumen depreciation (@40% rated life).

TYPICAL PERFORMANCE VALUES FOR COMPACT SOURCES *(cont.)*
(non-directional sources)

Lamp Types	Lamp Watts	System Watts	Initial Lamp CRI	Maintained Lamp Lumens	Maintained System Lumens	System Efficacy	Rated Life
Integral Units/Electronic							
Open Quad Tube	–	26	82	1500	1275	49	10000
Triple Twin/U-Tube	–	15	82	900	750	50	10000
Triple Twin/U-Tube	–	20	82	1200	1020	51	10000
Triple Twin/U-Tube	–	25	82	1520	1290	52	10000
Quadruple Twin-Tube	–	28	82	1750	1475	53	10000
T-4 Twin-Tube/Preheat Magnetic							
4.00-inch	5	9	82	225	216	24	10000
5.25-inch	7	11	82	360	324	29	10000
6.50-inch	9	13	82	540	432	33	10000
7.50-inch	13	17	82	810	792	47	10000
T-4 Quad Tube/Preheat Magnetic							
4.50-inch	9	13	82	540	440	34	10000
6.00-inch	13	17	82	774	756	44	10000

NOTES: Lamp lumen performance varies among manufacturers.
Maintained performance includes effect of lamp lumen depreciation (@40% rated life).

TYPICAL PERFORMANCE VALUES FOR COMPACT SOURCES *(cont.)*

(non-directional sources)

Lamp Types	Lamp Watts	System Watts	Initial Lamp CRI	Maintained Lamp Lumens	Maintained System Lumens	System Efficacy	Rated Life
T-4 Quad Tube/Preheat Magnetic							
7.00-inch	18	25	82	1125	1070	43	10000
7.50-inch	26	31	82	1620	1540	50	10000
Circline/Electronic Adapter							
6.1" diam.	–	13	84	950	800	62	12000
6.4" diam.	–	20	84	1450	1250	63	12000
8.2" diam.	–	22	84	1750	1475	67	12000
8.9" diam.	–	30	84	2400	2050	68	12000
2-D/Electronic Adapter							
5.5" diam.	–	22	82	1300	1100	50	10000
8.0" diam.	–	39	82	2780	2375	61	10000

NOTES: Lamp lumen performance varies among manufacturers.
Maintained performance includes effect of lamp lumen depreciation (@40% rated life).

TYPICAL PERFORMANCE VALUES FOR DIRECTIONAL LAMPS

Lamp Types	System Watts	Avg. System Lumens	CBCP Candelas	Beam Degrees	Rated Life
Incandescents					
R30	45	485	N/A	N/A	2000
R30/Krypton	60	775	510	65	2000
R30	75	900	470	72	2000
ER30	50	replaces 100W in deep cans			2000
ER30	75	replaces 150W in deep cans			2000
ER40	120	replaces 250W in deep cans			2000
R40	75	890	N/A	N/A	2000
R40	100	1190	N/A	N/A	2000
PAR38/ES/Spot	65	675	5900	14	2000
PAR38/ES/Flood	65	675	1750	30	2000
PAR38/Spot	75	765	4400	17	2000
PAR38/Flood	75	765	1750	33	2000
PAR38/ES/Spot	85	930	6800	15	2000

Notes: CBCP = center beam candlepower, the maximum luminous intensity (in candelas).
Beam angle = the angle in which the luminous intensity is at least 50% of the maximum value.

TYPICAL PERFORMANCE VALUES FOR DIRECTIONAL LAMPS *(cont.)*

Lamp Types	System Watts	Avg. System Lumens	CBCP Candelas	Beam Degrees	Rated Life
Incandescents					
PAR38/ES/Flood	85	930	2000	37	2000
PAR38/ES/Spot	120	1370	9200	18	2000
PAR38/ES/Flood	120	1370	3600	30	2000
PAR38/Spot	150	1740	12000	16	2000
PAR38/Flood	150	1740	3100	36	2000
Compact Halogen					
PAR30/Wide Flood	50	670	800	55	2000
PAR30/Spot/IR	50	1000	19500	7	3000
PAR30/Flood/IR	50	1000	2400	33	3000
PAR30/Spot	75	1100	15000	11	2000
PAR30/Flood	75	1100	2500	36	2000
PAR30/Wide Flood	75	1100	2500	36	2000
PAR38/Spot/IR	60	1150	18500	10	3000

Notes: CBCP = center beam candlepower, the maximum luminous intensity (in candelas).
Beam angle = the angle in which the luminous intensity is at least 50% of the maximum value.

TYPICAL PERFORMANCE VALUES FOR DIRECTIONAL LAMPS *(cont.)*

Lamp Types	System Watts	Avg. System Lumens	CBCP Candelas	Beam Degrees	Rated Life
Compact Halogen					
PAR38/Flood/IR	60	1150	3650	29	3000
PAR38/Spot	75	1070	18400	8	2500
PAR38/Flood	75	1070	4000	26	2500
PAR38/Spot	90	1270	18500	10	2000
PAR38/Flood	90	1270	4000	30	2000
PAR38/Wide Flood	90	1270	1500	55	2000
PAR38/Spot/IR	100	2000	30000	10	3000
PAR38/Flood/IR	100	2000	5500	33	3000
Compact Fluorescent					
Integral/Reflector/Quad	15	540	315	70	10000
Integral/Reflector/Quad	17	720	N/A	N/A	10000
Integral/Reflector/Quad	20	810	335	80	10000

Notes: CBCP = center beam candlepower, the maximum luminous intensity (in candelas).
Beam angle = the angle in which the luminous intensity is at least 50% of the maximum value.

FLOODLIGHT BEAM SPREADS

Beam Spread (degrees)	Lamp Type
10 – 18	1
18 – 29	2
29 – 46	3
46 – 70	4
70 – 100	5
100 – 130	6
130 and up	7

INCANDESCENT LAMP TROUBLESHOOTING GUIDE

Problem	Possible Cause	Corrective Action
Short lamp life	Voltage higher than lamp rating	Ensure that voltage is equal to or less than lamp's rated voltage. Lamp life is decreased when the voltage is higher than the rated voltage. A 5% higher voltage shortens the lamp life by 40%.
	Lamp exposed to rough service conditions or vibration	Replace lamp with one rated for rough service or resistant to vibration.
Lamp does not turn ON after new lamp is installed	Fuse blown, poor electrical connection, faulty control switch	Check circuit fuse or CB. Check electrical connections and voltage out of control switch. Replace switch when voltage is present into, but not out of, the control switch.

CHAPTER 5
LAMP DISPOSAL

Upgrading a lighting system almost always involves the removal and disposal of lamps and ballasts. Some of this waste may be considered hazardous, and you must manage it accordingly. Your company may face severe penalties if you do not.

PCB-CONTAINING BALLASTS

The primary concern regarding the disposal of used fluorescent ballasts is the health risk associated with polychlorinated biphenyls (PCBs), the same chemical that necessitated the removal of many oil-filled transformers in recent years. Fluorescent and HID ballasts use a small capacitor that may contain high concentrations of PCBs. The Toxic Substances Control Act (TSCA), enacted in 1976, subsequently banned the production of PCBs in the United States.

The proper method for disposing used ballasts depends on several factors, such as the type and condition of the ballasts and the regulations or recommendations in effect in the state(s) where you remove or discard them. TSCA specifies the disposal method for ballasts that are *leaking* PCBs. In addition, generators of PCB-containing ballast wastes may be subject to notification and liability provisions under the Comprehensive Environmental Response, Compensation and Liability Act of 1980 (CERCLA) also known as "Superfund." Because disposal requirements vary from state to state, check with regional, state or local authorities for all applicable regulations in your area.

IDENTIFYING PCB BALLASTS

Use the following guidelines to identify ballasts that contain PCBs:

- All ballasts manufactured through 1979 contain PCBs.
- Ballasts manufactured after 1979 that do not contain PCBs are labeled "No PCBs."

- If a ballast is not labeled "No PCBs," assume it contains PCBs.
- It is extremely important to find out if a ballast containing PCBs is leaking before you remove it from the fixture, so that you can handle it properly. If you do not, it can get very expensive.

NON-LEAKING PCB BALLAST DISPOSAL

Under federal laws, intact fluorescent and HID ballasts that are *not* leaking PCBs may be disposed in a municipal solid waste landfill. EPA recommends, but cannot demand, packing and sealing the intact ballasts in 55 gallon drums. Green Lights recommends high-temperature incineration, recycling or disposal in a chemical or hazardous waste landfill.

In addition, the federal laws regulate the disposal of non-leaking PCB-containing ballasts, requiring building owners and waste generators to notify the National Response Center at (800) 424-8802. They must notify when disposing a pound or more of PCBs (roughly equivalent to 12-16 fluorescent ballasts) in a 24-hour period.

A note of warning: The federal government has a history of applying some environmental laws after the fact (ex post facto), so as a generator of PCB-containing ballast wastes, you could be liable in a subsequent cleanup at a municipal, hazardous or chemical land disposal site, incinerator or recycling facility.

LEAKING PCB BALLASTS

A puncture or other damage to ballasts in a lighting system exposes an oily tar-like substance. If this substance contains PCBs, the ballast and all materials it contacts are considered PCB waste and are subject to federal requirements. Leaking PCB-containing ballasts must be incinerated at an EPA-approved high-temperature incinerator. Take precautions to prevent exposure of the leaking ballast, since all materials that contact the ballast or the leaking substance are also PCB waste. Use trained personnel or contractors to handle and dispose leaking

PCB-containing ballasts. For proper packing, storage, transportation and disposal information, call the TSCA assistance information hotline at (202) 554-1404.

STATE REQUIREMENTS

Many states have developed regulations governing the disposal of PCB-containing ballasts that are more stringent than Federal regulations. In addition, some EPA Regional offices have published policies specifying ballast disposal methods adopted by individual states.

State standards can take several forms (written regulations, regional policies, written and verbal recommendations, transportation documentation). Some states do not regulate PCB-containing ballasts as toxic waste, but prohibit their disposal in municipal solid waste landfills.

All generators of PCB-containing ballasts should thoroughly investigate their state's regulations and follow local requirements. There are three common methods for disposing of non-leaking PCB-containing ballasts: high-temperature incineration, recycling and chemical or hazardous waste landfill.

HIGH-TEMPERATURE INCINERATION

High-temperature incineration is the method preferred by many companies because it destroys PCBs, removing them from the waste stream permanently and removing the potential for future liability. Incinerating a PCB-containing ballast costs more than sending it to a hazardous waste landfill, but this additional cost is one many organizations are willing to absorb.

RECYCLING BALLASTS

Recyclers remove the PCB-containing materials (the capacitor and possibly the asphalt potting material surrounding the capacitor) for incineration or land disposal. Metals, such as copper and steel, can be reclaimed from the ballasts for use in manufacturing other products.

You may recycle used non-leaking ballasts despite PCBs.

CHEMICAL OR HAZARDOUS WASTE LANDFILL

PCB-containing ballasts may also be disposed in a chemical or hazardous waste landfill. Landfill disposal is less expensive than high-temperature incineration or recycling, but does not eliminate PCBs from the waste stream permanently. While chemical or hazardous waste landfill disposal is an acceptable, regulated disposal method, you may be legitimately concerned about potential future liability using this method.

PACKING PCB BALLASTS FOR DISPOSAL

Despite the disposal method selected, ballasts are packed according to PCB regulations in 55-gallon drums for transportation. One drum holds 150 to 300 ballasts depending on how tightly the ballasts are packed. You should fill the void spaces with an absorbent packing material for safety reasons, and label the drums according to Department of Transportation regulations.

Note that tightly packed drums may weigh more than 1,000 pounds, which may present a safety risk, particularly when moving the drum for loading or unloading.

DISPOSAL COSTS

High-temperature incineration and chemical or hazardous waste landfill costs can vary considerably. Disposal prices vary according to:

- Quantity of waste generated
- Location of removal site
- Proximity to an EPA-approved high-temperature incinerator or chemical or hazardous waste landfill
- State and local taxes

When shopping for ballast disposal services, request cost estimates in terms of both pounds and number of ballasts. Typical F40 ballasts weigh about 3.5 lbs., and F96 ballasts weigh about 8 lbs. Negotiate with hazardous waste brokers, transporters, waste management companies and disposal sites to obtain the lowest fees.

Note that these are average price figures, and you should verify prices in your area before estimating your costs.

HIGH-TEMPERATURE INCINERATION

Incineration costs are calculated by weight. Costs range from $0.55/lb. to $2.10/lb. Average cost is $1.50/lb., which equals approximately $5.25 per ballast. Note: Estimated costs do not include packaging, transportation or profile fees.

RECYCLING COSTS

When recyclers remove the PCB-containing capacitor, the volume and weight of the ballast are reduced. This change results in lower packing, transportation and incineration or disposal costs. Recycling costs are calculated by weight. Costs range from $0.75/lb. to $1.75/lb. Average cost is $1.00/lb., which equals approximately $3.50 per ballast. Note: Recycling cost can range from $1.25 per ballast (if the PCB wastes are sent to a chemical or hazardous waste landfill) to approximately $3.50 per ballast (if the PCB wastes are high-temperature incinerated). Estimated costs do not include packaging, transportation, or profile fees.

CHEMICAL OR HAZARDOUS WASTE LANDFILL

Chemical or hazardous waste landfill costs are calculated per 55-gallon drum. Costs range from $65/drum to $165/drum. Average cost is $100/drum, which equals approximately $0.50/ballast. Note: Estimated costs do not include packaging, transportation or profile fees.

TRANSPORTATION

Transportation fees are calculated as cents per pound per mile. They vary according to (1) the number of drums removed from the site, and (2) the distance from your location to the location of the high-temperature incinerator, chemical or hazardous waste landfill, or recycler.

Transporters may need to be registered or licensed to move hazardous wastes in certain states. Documentation of the movement of hazardous waste may be required even if a state does not regulate disposal or fees require the use of a licensed transporter.

PROFILE FEES

Operators of the high-temperature incinerator or chemical or hazardous waste landfill may charge a profile fee to document incoming hazardous waste. Profile fees vary depending on the volume of waste materials generated. Profile fees range from $0 to $300 per delivery.

Fees may be waived if a certain volume or frequency of deliveries is assured or a working relationship has been established with a waste management broker, lighting management company, or other contractor.

RECORD-KEEPING

To track transported hazardous waste, EPA requires generators to prepare a Uniform Hazardous Waste Manifest. The hazardous waste landfill, incinerator, or recycler that you use can provide this one-page form. The manifest identifies the type and quantity of waste, the generator, the transporter and its ultimate destination.

The manifest must accompany the waste wherever it travels. Each handler of the waste must sign the manifest and keep one copy. When the waste reaches its destination, the owner of that facility returns a copy of the manifest to the generator to confirm that the waste arrived. If the waste does not arrive as scheduled, generators must immediately notify EPA or the authorized state environmental agency, so that they can investigate and act appropriately. In addition, require your contractor to provide you with documents verifying the disposal method, whether the PCBs are incinerated at high-temperatures or disposed in a chemical or hazardous waste landfill.

DISPOSAL OF MERCURY-CONTAINING LAMPS

Fluorescent and high-intensity discharge (HID) lamps contain a small quantity of mercury that can be harmful to the environment and to human health when improperly managed. Mercury is regulated under regulations administered by the U.S. Environmental Protection Agency. Under current federal law, mercury-containing lamps such as fluorescent and HID lamps may or may not be considered hazardous waste. In addition, incandescent and HID lamps may contain small quantities of lead that can also be potentially harmful to human health and the environment.

The federal government requires generators of solid wastes containing mercury to determine whether or not the waste is hazardous by using generator knowledge or testing representative samples of that waste. According to the regulations, generators of used fluorescent and HID lamps are responsible for determining whether their lamp wastes are hazardous. If you do not test used fluorescent and HID lamps and prove them non-hazardous, assume they are hazardous waste and dispose them accordingly.

To use generator knowledge in making a hazardous waste determination, the generator must have information on possible hazardous constituents and their quantities in the waste. Sometimes manufacturers generate solid waste as part of their manufacturing process, and can use process knowledge to determine whether the waste exhibits a characteristic of hazardous waste. However, with expired lamp wastes the generator has little process knowledge on which to make a hazardous waste determination (since he is not the manufacturer). The generator could base a determination on data obtained from the manufacturer or he could refer to EPA's study entitled "Analytical Results of Mercury in Fluorescent Lamps."

TESTING LAMPS

A test called the Toxicity Characteristic Leaching Procedure (TCLP) identifies whether a waste is toxic and must be managed as hazardous waste. The test attempts to replicate the conditions in a municipal landfill to detect the mercury concentration of water that would leach from the landfill.

When mercury-containing lamps are tested using the TCLP, the test results can vary considerably, depending on the lamp manufacturer, the age of the lamp and the laboratory procedures used. These lamps often fail the TCLP. If you do not use the TCLP to verify that your lamps are non-hazardous, you should (1) assume that they are hazardous waste, and (2) manage them as hazardous waste. Contact your state hazardous waste agency for information on laboratories in your state that conduct the TCLP test. The cost to test one lamp is usually about $140.

There is exemption from some disposal requirements if you or your customers qualify as a *conditionally exempt small quantity generator*, which is defined as a generator who disposes 100 kg or less of hazardous waste per month. Generators must add the weight of all the hazardous waste (lamps plus other hazardous wastes) that their business generates during a month. For lamp disposal, this quantity of waste includes the mercury in the lamp along with the glass, phosphors and other materials (the weight of the entire lamp). Conditionally exempt small quantity generators are excused from identification, storage, treatment and disposal regulations. To qualify as a conditionally exempt small quantity generator (if the only hazardous waste is mercury-containing lamps), a generator must dispose of fewer than 300-350 four-foot T12 fluorescent lamps or 400-450 four-foot T8 fluorescent lamps per month, depending upon the approximate weight of each lamp.

RECYCLING

Recycling costs for fluorescent lamps are typically calculated by linear foot. HID lamp recycling costs are typically quoted on a per-lamp basis. Fluorescent recycling costs range from $0.06/ft. to $0.15/ft., with the average cost being $0.10/ft. Approximately $0.40 per F40 lamp. HID recycling costs range from $1.25/lamp to $4.50/lamp, average cost is $2.50/lamp. Note: Estimated costs do not include packaging, transportation or profile fees.

CHEMICAL OR HAZARDOUS WASTE LANDFILL

Disposal costs for fluorescent lamps at a hazardous waste landfill range from 25-50 cents per 4-foot tube, not including costs for packaging, transportation or profile fees.

PACKING LAMPS

To prevent used fluorescent and HID lamps from breaking, lamps should be properly packed for storage and transportation. When lamps are removed and replaced with new lamps (during group relamping), the used lamps should be packed in the cardboard boxes that contained the replacement lamps. The boxes containing the hazardous waste must be properly labeled.

Pre-printed labels or rubber stamps that meet Department of Transportation regulations are recommended for high-volume disposal.

Small quantity generators dispose 100 to 1,000 kg of hazardous waste per month (which roughly corresponds to 350 to 3,600 four-foot lamps) and can store hazardous waste up to 180 days.

Large quantity generators dispose over 1,000 kg of hazardous waste per month (more than 3,600 four-foot lamps) and can store hazardous waste up to 90 days.

Conditionally exempt small quantity generators dispose 100 kg or less of hazardous waste per month and are exempt from storage requirements. In addition to proper packing, care should be taken when stacking the boxes of used lamps for storage to avoid crushing the bottom boxes under the weight of the boxes on top. If you work with a contractor to maintain your lighting system, you may want to specify a safe storage arrangement in your contract. This approach ensures that your used lamps are not accidentally broken or crushed before they are sent to a disposal facility.

Some organizations crush their used lamps before disposal. This option should be pursued with care. The crushing equipment should have the approval of state and local authorities, and crushing methods should be evaluated carefully. The lamp should be crushed entirely inside the drum or storage unit so that no mercury vapor enters the atmosphere. There should also be adequate ventilation in the space where the crushing occurs. Under current EPA hazardous waste regulations, crushing lamps before sending them to a hazardous waste landfill may be considered treatment. A treatment permit may be required.

DECISION CHART

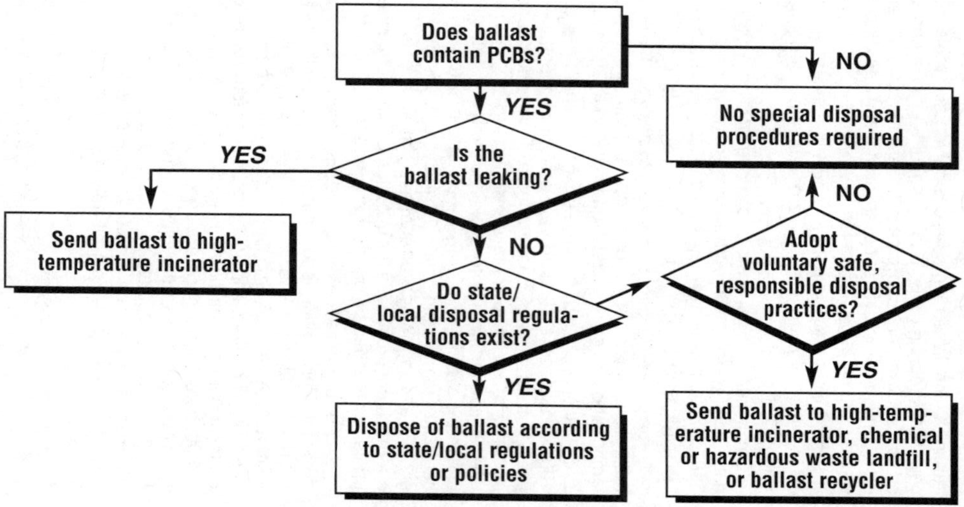

INFORMATIONAL RESOURCES

Postal Addresses for the U.S. Environmental Protection Agency

To contact individuals and offices within the U.S. Environmental Protection Agency, use the following addresses and mail codes. To find an individual's mail code, use the EPA Employee Directory.

EPA Headquarters

Standard Mailing Address
Environmental Protection Agency
Ariel Rios Building
1200 Pennsylvania Avenue, N.W.
Mail Code 3213A
Washington, DC 20460
(202) 260-2090

Overnight Package Delivery Mailing Address
Environmental Protection Agency
EPA East
1201 Constitution Avenue, N.W.
Room number 4101 M
Washington, DC 20004

EPA Regional Offices

REGION 1 (CT, MA, ME, NH, RI, VT)
Environmental Protection Agency
1 Congress Street, Suite 1100
Boston, MA 02114-2023
http://www.epa.gov/region01/
Phone: (617) 918-1111
Toll free within Region 1: (888) 372-7341
Fax: (617) 565-3660

REGION 2 (NJ, NY, PR, VI)
Environmental Protection Agency
290 Broadway
New York, NY 10007-1866
http://www.epa.gov/region02/
Phone: (212) 637-3000
Fax: (212) 637-3526

REGION 3 (DC, DE, MD, PA, VA, WV)
Environmental Protection Agency
1650 Arch Street
Philadelphia, PA 19103-2029
http://www.epa.gov/region03/
Phone: (215) 814-5000
Toll free: (800) 438-2474
Fax: (215) 814-5103
Email: r3public@epa.gov

REGION 4 (AL, FL, GA, KY, MS, NC, SC, TN)
Environmental Protection Agency
Atlanta Federal Center
61 Forsyth Street, SW
Atlanta, GA 30303-3104
http://www.epa.gov/region04/
Phone: (404) 562-9900
Toll free: (800) 241-1754
Fax: (404) 562-8174

REGION 5 (IL, IN, MI, MN, OH, WI)
Environmental Protection Agency
77 West Jackson Boulevard
Chicago, IL 60604-3507
http://www.epa.gov/region05/
Phone: (312) 353-2000
Toll free within Region 5: (800) 621-8431
Fax: (312) 353-4135

REGION 6 (AR, LA, NM, OK, TX)
Environmental Protection Agency
Fountain Place 12th Floor, Suite 1200
1445 Ross Avenue
Dallas, TX 75202-2733
http://www.epa.gov/region06/
Phone: (214) 665-2200
Toll free within Region 6: (800) 887-6063
Fax: (214) 665-7113

REGION 7 (IA, KS, MO, NE)
Environmental Protection Agency
901 North 5th Street
Kansas City, KS 66101
http://www.epa.gov/region07/
Phone: (913) 551-7003
Toll free: (800) 223-0425

REGION 8 (CO, MT, ND, SD, UT, WY)
Environmental Protection Agency
999 18th Street, Suite 500
Denver, CO 80202-2466
http://www.epa.gov/region08/
Phone: (303) 312-6312
Toll free: (800) 227-8917
Fax: (303) 312-6339
Email: r8eisc@epa.gov

REGION 9 (AZ, CA, HI, NV)
Environmental Protection Agency
75 Hawthorne Street
San Francisco, CA 94105
http://www.epa.gov/region09/
Phone: (415) 947-8000
Toll free within Region 9: (866) EPA-WEST
Fax: (415) 947-3553
Email: r9.info@epa.gov

REGION 10 (AK, ID, OR, WA)
Environmental Protection Agency
1200 Sixth Avenue
Seattle, WA 98101
http://www.epa.gov/region10/
Phone: (206) 553-1200
Toll free: (800) 424-4372
Fax: (206) 553-0149

Solid and Hazardous Waste Agencies per State

ALABAMA
Department of Environmental Management
Land Division – Solid/Hazardous Waste
P.O. Box 301463
Montgomery, AL 36130-1463
(334) 271-7730

ALASKA
State of Alaska
Department of Environmental Conservation
410 Willoughby Avenue, Suite 303
Juneau, AK 99801-1795
(907) 465-5010

ARIZONA
Arizona Department of Environmental Quality
Hazardous Waste Permits Unit
1110 W. Washington Street
Phoenix, AZ 85007
(602) 771-4153

ARKANSAS
Department of Environmental Quality
Hazardous Waste Division
8001 National Drive
Little Rock, AR 72209
(501) 682-0833

CALIFORNIA
Department of Toxic Substances Control
P.O. Box 806
Sacramento, CA 95812-0806
(800) 728-6942

COLORADO
Hazardous Materials and Waste Management Division
Colorado Department of Public Health and Environment
Mail Code: HMWMD-HWC-B2
4300 Cherry Creek Drive South
Denver, CO 80222-1530
(888) 569-1831

CONNECTICUT
Department of Environmental Protection
Waste Management Bureau
79 Elm Street
Hartford, CT 06106-5127
(860) 424-4081

DELAWARE
Department of Natural Resources
and Environmental Control
Division of Environmental Control
Solid Waste/Hazardous Waste Section
89 Kings Highway
Dover, DE 19901
(302) 739-3689

DISTRICT OF COLUMBIA
Hazardous Waste Environmental Health Administration
51 N Street, 3rd Floor
Washington, DC 20002
(202) 535-2270

FLORIDA
Bureau of Solid and Hazardous Waste
Department of Environmental Protection
3900 Commonwealth Boulevard, M.S. 49
Tallahassee, FL 32399
(850) 245-8707

GEORGIA
Environmental Protection Division
Hazardous Waste Program
2 Martin Luther King Drive SE
Atlanta, GA 30334
(404) 656-7802

HAWAII
State of Hawaii
Department of Health
Environmental Management Division
Solid & Hazardous Waste Branch
P.O. Box 3378
Honolulu, HI 96801-3378
(808) 586-4226

IDAHO
Department of Health and Welfare
Division of Environment
Bureau of Hazardous Materials
4040 Guard Street
Boise, ID 83705-5004
(208) 422-5725

ILLINOIS
State of Illinois
Environmental Protection Agency
Bureau of Land
1021 N Grand Avenue East
Springfield, IL 62702
(217) 524-3300

INDIANA
Indiana Dept. of Environmental Management
Industrial Waste Permit Section
P.O. Box 6015
Indianapolis, IN 46206-6015
(317) 308-3003

IOWA
Department of Natural Resources
Solid Waste Section
Land Quality Bureau
502 E. 9th Street
Des Moines, IA 50319-0034
(515) 281-8986

KANSAS
Department of Health and Environment
Bureau of Waste Management
1000 W. Jackson Street, Suite 320
Topeka, KS 66612-1366
(785) 296-1601

KENTUCKY
NREPC
Division of Waste Management
Ft. Boone Plaza
14 Reilly Road
Frankfort, KY 40601
(502) 564-6716

LOUISIANA
Department of Environmental Quality
Office of Environmental Assessment
Solid Waste Division
P.O. Box 4314
Baton Rouge, LA 70821-4314
(225) 765-0355

MAINE
Department of Environmental Protection
Remediation and Waste Management
17 State House Station
Augusta, ME 04333-0017
(207) 287-7688

MARYLAND
Department of Environment
Hazardous Waste Program
1800 Washington Boulevard
Baltimore, MD 21230
(410) 537-3345

MASSACHUSETTS
Department of Environmental Protection
Office of Hazardous Waste
1 Winter Street
Boston, MA 02108
(617) 292-5898

MICHIGAN
Department of Natural Resources
Waste and Hazardous Materials Division
PO Box 30241
Lansing, MI 48909
(517) 373-2730

MINNESOTA
Minnesota Pollution Control Agency
Solid or Hazardous Waste Division
520 Lafayette Road North
St. Paul, MN 55155-4194
(651) 297-2274

MISSISSIPPI
Department of Environmental Quality
Office of Pollution Control
PO Box 20305
Jackson, MS 39289
(601) 961-5100

MISSOURI
Department of Natural Resources
Division of Environmental Quality
Hazardous Waste Program
Jefferson State Office Building
P.O. Box 176
Jefferson City, MO 65102
(573) 751-3176

MONTANA
Department of Environmental Quality
Permitting/Compliance Division
Air and Waste Management Division
P.O. Box 200901
Helena, MT 59620-0901
(406) 444-3490

NEBRASKA
Department of Environmental Quality
1200 "N" Street, Suite 400
P.O. Box 98922
State Office Building
Lincoln, NE 68509
(402) 471-2186

NEVADA
Bureau of Management Waste
Hazardous Waste Management Program
333 West Nye Lane
Carson City, NV 89706-0851
(775) 687-4670

NEW HAMPSHIRE
Department of Environmental Services
Waste Management Division/
Hazardous Waste Compliance
6 Hazen Drive
Concord, NH 03301
(603) 271-2942

NEW JERSEY
New Jersey Department of Environmental Protection
Division of Solid and Hazardous Waste
401 East State Street
P.O. Box 414
Trenton, NJ 08625-0414
(609) 633-1418

NEW MEXICO
New Mexico Environment Department
Hazardous Waste Bureau
2905 Rodeo Park Drive East
Building 1
Santa Fe, NM 87505
(505) 428-2500

NEW YORK
Division of Solid and Hazardous Materials
New York State Department of
Environmental Conservation
625 Broadway
Albany, NY 12233
(518) 402-8712

NORTH CAROLINA
Department of Environment, Health,
and Natural Resources
Solid Waste Management/Hazardous Waste Division
1601 Mail Service Center
Raleigh, NC 27699-1601
(919) 733-4996

NORTH DAKOTA
Health Department, Environmental Health Section
Division of Waste Management
1200 Missouri Avenue
P.O. Box 5520
Bismarck, ND 58506-5520
(701) 328-5166

OHIO
Environmental Protection Agency
Office of Solid and Infectious Management
122 S. Front Street
Columbus, OH 43215
(614) 644-3020

OKLAHOMA
Oklahoma Department of Environmental Quality
Land Protection Division
P.O. Box 1677
Oklahoma City, OK 73101-1677
(405) 702-5100

OREGON
Department of Environmental Quality
Waste Management Clean-up Division
811 S.W. 6th Avenue
Portland, OR 97204-1390
(503) 229-5913

PENNSYLVANIA
Department of Environmental Protection
Division of Hazardous Waste Management
PO Box 8471
Harrisburg, PA 17105-8471
(717) 787-6239

PUERTO RICO
Environmental Quality Board
Solid and Hazardous Waste Bureau
National Plaza Building, 12th Floor
Ponce de Leon Avenue, #431
Hato Rey, PR 00917
(787) 767-8056

RHODE ISLAND
Department of Environmental Management
Office of Waste Management
235 Promenade Street
Providence, RI 02908-5767
(401) 277-2797

SOUTH CAROLINA
Board of Health and Environmental Control
Bureau of Land and Waste Management
2600 Bull Street
Columbia, SC 29201
(803) 896-4000

SOUTH DAKOTA
Department of Water and Natural Resources
Environmental Health Division
Waste Management Program
Joe Foss Building
523 E. Capitol
Pierre, SD 57501
(605) 773-3153

TENNESSEE
Department of Environment and Conservation
Division of Solid Waste Management
5th Floor, L&C Tower
401 Church Street
Nashville, TN 37243-1535
(615) 532-0780

TEXAS
Texas Commission on Environmental Quality
P.O. Box 13087
Austin, TX 78711-3087
(512) 239-2334

UTAH
Department of Environmental Quality
Division of Solid and Hazardous Waste
168 North 1950 West
Salt Lake City, UT 84116
(801) 538-6765

VERMONT
Department of Environmental Conservation
Environmental Assistance Division
103 South Main Street
Landry Building
Waterbury, VT 05671-0411
(802) 241-3589

VIRGINIA
Virginia Department of Environmental Quality
Waste Management
P.O. Box 10009
Richmond, VA 22240
(804) 698-4193

WASHINGTON
Department of Ecology
Solid and Hazardous Waste Program
PO Box 47600
Olympia, WA 98504-7600
(206) 407-6702

WEST VIRGINIA
West Virginia Division of Environmental Protection
Division of Water and Waste Management
1356 Hansford Street
Charleston, WV 25301
(304) 558-5989

WISCONSIN
Department of Natural Resources
Division of Air and Waste
P.O. Box 7921
101 South Webster Street
Madison, WI 53707-7921
(608) 266-2621

WYOMING
Department of Environmental Quality
Solid Waste Management Program
122 West 25th Street
Herscheler Building
Cheyenne, WY 82002
(307) 777-7752

TSCA, RCRA, and CERCLA Information Phone Lines

Toxic Substances Control Act (TSCA)
Assistance Information Hotline
(202) 554-1404

RCRA/CERCLA Hotline
(800) 424-9346
in the Washington, DC Metro Area
(703) 412-9810

CERCLA National Response Center
(NRC) Hotline
(800) 424-8802

EPA-Approved Disposal Locations

Commercially permitted PCB INCINERATORS operating as of June 1993

Aptus, Inc.
PO Box 1328
Coffeyville, KS 67337
(316) 251-6380

Aptus, Inc.
Aragonite, UT
(801) 266-7787

Chemical Waste Management
PO Box 2563
Port Arthur, TX 77643
(409) 736-2821

Environmental Energy Group
Denton, TX
(817) 383-3632

Environmental Energy Group
PO Box 50764
Denton, TX 76206
(817) 898-1291

Rollins
PO Box 609
Deer Park, TX 77536
(713) 930-2300

Commercially permitted HAZARDOUS WASTE LANDFILLS operating as of June 1993

Chem-Security Systems Incorporated
Star Route, Box 9
Arlington, OR 98712
(503) 454-2643

Chemical Waste Management
Call 1-800-843-3604
for information on CWM disposal facilities nationwide.

Envirosafe Services Inc. of Idaho
PO Box 16217
Boise, ID 83715-6217
(800) 274-1516

US Ecology, Inc.
Box 578
Beatty, NV 89003
(702) 553-2203

US Pollution Control, Inc.
Grayback Mountain
8960N Hwy 40
Lake Point, UT 84074
(801) 531-4980

Recycling Resources

Lamp Recycling Services

Advanced Environmental Recycling Corp.
2591 Mitchell Avenue
Allentown, PA
(800) 554-2372 or (215) 797-7608

Allied Technology Group
47375 Freemont Boulevard
Freemont, CA 94538
(510) 490-3008

Alta Resource Management Services
88-B Industry Avenue
Springfield, MA 01104-9926
(800) 730-ALTA or (413) 734-3399

Bethlehem Apparatus
Hellertown, PA
(215) 838-7034

Dynex Environmental, Inc.
6801 Industrial Loop
Milwaukee, WI 53129
(800) 249-3310 or (414) 421-4959
4751 Mustang Circle
St. Paul, MN 55112
(800) 733-9639 or (612) 784-4040

Global Recycling Technologies, Inc.
P.O. Box 651
Randolph, MA 02368
(617) 341-6080

Light Cycle, Inc.
1222 University Avenue
St. Paul, MN 55104
(612) 641-1309

Lighting Resources, Inc.
386 S. Gordon Street
Pomona, CA
(800) 57-CYCLE

Luminaire Recyclers Inc.
2161 University Avenue, Suite 206
St. Paul, MN 55114
(612) 649-0079

Mercury Recovery Systems
2021 S. Myrtle Street
Monrovia, CA
(818) 301-1372

Mercury Refining Co., Inc.
Albany, NY
(518) 459-0820

Mercury Technologies International, LP
Hayward, CA
(800) 628-3675
Los Angeles, CA
(310) 475-4684
West Melbourne, FL
(407) 852-1516

Mercury Technologies of Minnesota
Pine City Industrial Park
Pine City, MN 55063-0013
(612) 629-7888
(800) 864-3821

Nine West Technologies
Nashville, TN
(615) 399-1486

NSSI, Inc.
574 Etheridge Street
Houston, Texas 77087

Recycle Technologies, Inc.
1480 N. Springdale Road
Waukesha, WI 53186
(800) 305-3040
(414) 798-3040

Recyclights
2010 E. Hennepin Avenue
Minneapolis, MN 55413
(612) 378-9571

Resource Recovery, Inc.
Edina, MN
(612) 828-9722 (service)
(701) 234-9102 (sales)

Superior Lamp Recycling, Inc.
Mineral Springs Facility
1275 Mineral Springs Drive
Port Washington, WI 53074
(800) 556-LAMP (5267)

USA Lamp and Ballast Recyclers
Call John Fortino at 1-800-778-6645
for information on disposal facilities.

USA Lights
2007 County Road, C-2
Roseville, MN 55113
(612) 628-9370

Ballast Recycling Services

Alta Resource Management Services, Inc.
88-B Industry Avenue
Springfield, MA 01104-9926
(800) 730-ALTA or
(413) 734-3399

Dynex Environmental, Inc.
6801 Industrial Loop
Milwaukee, WI 53129
(800) 249-3310 or
(414) 421-4959

4751 Mustang Circle
St. Paul, MN 55112
(800) 733-9639 or
(612) 784-4040

Eastern Environmental Technologies
Portchester, NY
(914) 934-2100

Ensquare, Inc.
Newton Upper Falls, MA
(617) 776-7320

FulCircle Ballast Recyclers
168 Brattle Street
Cambridge, MA
(800) 775-1516

Baltimore, MD
(717) 932-1022

New York, NY
(800) 581-0857

San Francisco, CA
(916) 649-9194

Los Angeles, CA
(800) 775-1516

Atlanta, GA
(800) 775-1516

FulCircle Ballast Recyclers *(cont.)*

Chicago, IL
(708) 434-0593

Detroit, MI
(313) 651-6589

Global Recycling Technologies, Inc.
PO Box 651
Randolph, MA 02368
(617) 341-6080

Lighting Resources, Inc.
Pomona, CA
(714) 622-0881

Light Cycle, Inc.
1222 University Avenue
St. Paul, MN 55104
(612) 641-1309

Luminaire Recyclers Inc.
2161 University Avenue,
Suite 206
St. Paul, MN 55114
(612) 649-0079

Recycle Technologies, Inc.
1480 N. Springdale Road
Waukesha, WI 53186
(800) 305-3040
(414) 798-3040

S.D. Myers
180 South Avenue
Tallmadge, Ohio 44278
(216) 633-2666

Ballast Recycling Services *(cont.)*

Salesco U.S.A.
Boston, MA
(617) 344-4074
Chicago, IL
(708) 803-0880
Dallas, TX
(214) 661-8819
Honolulu, HI
(800) 368-9095
Phoenix, AZ
(800) 368-9095
San Diego, CA
(619) 793-3460

Transformer Service, Inc.
Concord, NH 03302
(603) 224-4006

Transtec Environmental
Niagara Falls, NY
(716) 283-6174

USA Lamp and Ballast Recyclers
Call John Fortino at
1-800-778-6645
for information on disposal facilities.

United States Ballast
Wausau, WI
(800) 715-5267

This is not a complete list of companies who provide recycling and disposal services throughout the United States. Companies listed in this section are not endorsed by the EPA or the green lights program. EPA does not screen listed companies and cannot confirm the methods these companies may use in their recycling process.

CHAPTER 6
LIGHTING TERMS

AMPERE: The standard unit of measurement for electric current that is equal to one coulomb per second. It defines the quantity of electrons moving past a given point in a circuit during a specific period. Amp is an abbreviation.

ANSI: Abbreviation for American National Standards Institute.

ARC TUBE: A tube enclosed by the outer glass envelope of a HID lamp and made of clear quartz or ceramic that contains the arc stream.

ASHRAE: American Society of Heating, Refrigerating and Air Conditioning Engineers.

BAFFLE: A single opaque or translucent element used to control light distribution at certain angles.

BALLAST: A device used to operate fluorescent and HID lamps. The ballast provides the necessary starting voltage, while limiting and regulating the lamp current during operation.

BALLAST CYCLING: Undesirable condition under which the ballast turns lamps on and off (cycles) due to the overheating of the thermal switch inside the ballast. This may be due to incorrect lamps, improper voltage being supplied, high ambient temperature around the fixture, or the early stage of ballast failure.

BALLAST EFFICIENCY FACTOR: The ballast efficiency factor (BEF) is the ballast factor (see below) divided by the input power of the ballast. The higher the BEF (within the same lamp-ballast type) the more efficient the ballast.

BALLAST FACTOR: The ballast factor (BF) for a specific lamp-ballast combination represents the percentage of the rated lamp lumens that will be produced by the combination.

CANDELA: Unit of luminous intensity, describing the intensity of a light source in a specific direction.

LIGHTING TERMS (cont.)

CANDELA DISTRIBUTION: A curve, often on polar coordinates, illustrating the variation of luminous intensity of a lamp or luminaire in a plane through the light center.

CANDLEPOWER: A measure of luminous intensity of a light source in a specific direction, measured in candelas (see above).

CBM: Abbreviation for Certified Ballast Manufacturers Association.

CEC: Abbreviation for California Energy Commission.

COEFFICIENT OF UTILIZATION: The ratio of lumens from a luminaire received on the work plane to the lumens produced by the lamps alone. (Also called "CU")

COLOR RENDERING INDEX (CRI): A scale of the effect of a light source on the color appearance of an object compared to its color appearance under a reference light source. Expressed on a scale of 1 to 100, where 100 indicates no color shift. A low CRI rating suggests that the colors of objects will appear unnatural under that particular light source.

COLOR TEMPERATURE: The color temperature is a specification of the color appearance of a light source, relating the color to a reference source heated to a particular temperature, measured by the thermal unit Kelvin. The measurement can also be described as the "warmth" or "coolness" of a light source. Generally, sources below 3200K are considered "warm"; while those above 4000K are considered "cool" sources.

COMPACT FLUORESCENT: A small fluorescent lamp that is often used as an alternative to incandescent lighting. The lamp life is about 10 times longer than incandescent lamps and is 3-4 times more efficacious. Also called PL, Twin-Tube, CFL, or BIAX lamps.

LIGHTING TERMS (cont.)

CONSTANT WATTAGE (CW) BALLAST: A premium type of HID ballast in which the primary and secondary coils are isolated. It is considered a high performance, high loss ballast featuring excellent output regulation.

CONSTANT WATTAGE AUTOTRANSFORMER (CWA) BALLAST: A popular type of HID ballast in which the primary and secondary coils are electrically connected. Considered an appropriate balance between cost and performance.

CONTRAST: The relationship between the luminance of an object and its background.

CRI: (SEE **COLOR RENDERING INDEX**)

CUT-OFF ANGLE: The angle from a fixture's vertical axis at which a reflector, louver, or other shielding device cuts off direct visibility of a lamp. It is the complementary angle of the shielding angle.

DAYLIGHT COMPENSATION: A dimming system controlled by a photocell that reduces the output of the lamps when daylight is present. As daylight levels increase, lamp intensity decreases. An energy-saving technique used in areas with significant daylight contribution.

DIFFUSE: Term describing dispersed light distribution. Refers to the scattering or softening of light.

DIFFUSER: A translucent piece of glass or plastic sheet that shields the light source in a fixture. The light transmitted throughout the diffuser will be redirected and scattered.

DIRECT GLARE: Glare produced by a direct view of light sources. Often the result of insufficiently shielded light sources. (See **GLARE**)

DOWNLIGHT: A type of ceiling luminaire, usually fully recessed, where most of the light is directed downward. May feature an open reflector and/or shielding device.

LIGHTING TERMS (cont.)

EFFICACY: A metric used to compare light output to energy consumption. Efficacy is measured in lumens per watt. Efficacy is similar to efficiency, but is expressed in dissimilar units. For example, if a 100-watt source produces 9000 lumens, then the efficacy is 90 lumens per watt.

ELECTROLUMINESCENT: A light source technology used in exit signs that provides uniform brightness, long lamp life (approximately eight years), while consuming very little energy (less than one watt per lamp).

ELECTRONIC BALLAST: A ballast that uses semiconductor components to increase the frequency of fluorescent lamp operation (typically in the 20-40 kHz range). Smaller inductive components provide the lamp current control. Fluorescent system efficiency is increased due to high frequency lamp operation.

ELECTRONIC DIMMING BALLAST: A variable output electronic fluorescent ballast.

EMI: Abbreviation for electromagnetic interference. High frequency interference (electrical noise) caused by electronic components or fluorescent lamps that interferes with the operation of electrical equipment. EMI is measured in microvolts, and can be controlled by filters. Because EMI can interfere with communication devices, the Federal Communication Commission (FCC) has established limits for EMI.

ENERGY-SAVING BALLAST: A type of magnetic ballast designed so that the components operate more efficiently, cooler and longer than a "standard magnetic" ballast. By U.S. law, standard magnetic ballasts can no longer be manufactured.

ENERGY-SAVING LAMP: A lower wattage lamp, generally producing fewer lumens.

FC: (SEE **FOOT-CANDLE**)

FLUORESCENT LAMP: A light source consisting of a tube filled with argon, along with krypton or other inert gas. When electrical current is applied, the resulting arc emits ultraviolet radiation that excites the phosphors inside the lamp wall, causing them to radiate visible light.

FOOT-CANDLE (FC): The English unit of measurement of the illuminance (or light level) on a surface. One foot-candle is equal to one lumen per square foot.

FOOTLAMBERT: English unit of luminance. One footlambert is equal to 1/p candelas per square foot.

GLARE: The effect of brightness or differences in brightness within the visual field sufficiently high to cause annoyance, discomfort or loss of visual performance.

HALOGEN: (SEE **TUNGSTEN HALOGEN LAMP**)

HARMONIC DISTORTION: A harmonic is a sinusoidal component of a periodic wave having a frequency that is a multiple of the fundamental frequency. Harmonic distortion from lighting equipment can interfere with other appliances and the operation of electric power networks. The total harmonic distortion (THD) is usually expressed as a percentage of the fundamental line current. THD for 4-foot fluorescent ballasts usually range from 20% to 40%. For compact fluorescent ballasts, THD levels greater than 50% are not uncommon.

HID: Abbreviation for high intensity discharge. Generic term describing mercury-vapor, metal-halide, high-pressure sodium, and (informally) low-pressure sodium light sources and luminaires.

HIGH-BAY: Pertains to the type of lighting in an industrial application where the ceiling is 20 feet or higher. Also describes the application itself.

HIGH OUTPUT (HO): A lamp or ballast designed to operate at higher currents (800 mA) and produce more light.

HIGH POWER FACTOR: A ballast with a 0.9 or higher rated power factor, which is achieved by using a capacitor.

HIGH-PRESSURE SODIUM LAMP: A high intensity discharge (HID) lamp whose light is produced by radiation from sodium vapor (and mercury).

HOT RESTART or HOT RESTRIKE: The phenomenon of restriking the arc in an HID light source after a momentary power loss. Hot restart occurs when the arc tube has cooled a sufficient amount.

IESNA: Abbreviation for Illuminating Engineering Society of North America.

ILLUMINANCE: A photometric term that quantifies light incident on a surface or plane. Illuminance is commonly called light level. It is expressed as lumens per square foot (foot-candles), or lumens per square meter (lux).

INDIRECT GLARE: Glare produced from a reflective surface.

INSTANT START: A fluorescent circuit that ignites the lamp instantly with a very high starting voltage from the ballast. Instant start lamps have single-pin bases.

LAMP CURRENT CREST FACTOR (LCCF): The peak lamp current divided by the RMS (average) lamp current. Lamp manufacturers require <1.7 for best lamp life. An LCCF of 1.414 is a perfect sine wave.

LAMP LUMEN DEPRECIATION FACTOR (LLD): A factor that represents the reduction of lumen output over time. The factor is commonly used as a multiplier to the initial lumen rating in illuminance calculations, which compensates for the lumen depreciation. The LLD factor is a dimensionless value between 0 and 1.

LAY-IN-TROFFER: A fluorescent fixture; usually a 2' x 4' fixture that sets or "lays" into a specific ceiling grid.

LIGHTING TERMS (cont.)

LED: Abbreviation for light emitting diode. An illumination technology used for exit signs. Consumes low wattage and has a rated life of greater than 80 years.

LENS: Transparent or translucent medium that alters the directional characteristics of light passing through it. Usually made of glass or acrylic.

LIGHT LOSS FACTOR (LLF): Factors that allow for a lighting system's operation at less than initial conditions. These factors are used to calculate maintained light levels. LLFs are divided into two categories, recoverable and nonrecoverable. Examples are lamp lumen depreciation and luminaire surface depreciation.

LIFE-CYCLE COST: The total costs associated with purchasing, operating, and maintaining a system over the life of that system.

LOUVER: Grid type of optical assembly used to control light distribution from a fixture. Can range from small-cell plastic to the large-cell anodized aluminum louvers used in parabolic fluorescent fixtures.

LOW POWER FACTOR: Essentially, an uncorrected ballast power factor of less than 0.9. (SEE **NPF**)

LOW-PRESSURE SODIUM: A low-pressure discharge lamp in which light is produced by radiation from sodium vapor. Considered a monochromatic light source (most colors are rendered as gray).

LOW-VOLTAGE LAMP: A lamp (typically compact halogen) that provides both intensity and good color rendition. Lamp operates at 12V and requires the use of a transformer. Popular lamps are MR11, MR16, and PAR36.

LOW-VOLTAGE SWITCH: A relay (magnetically-operated switch) that allows local and remote control of lights, including centralized time clock or computer control.

LIGHTING TERMS (cont.)

LUMEN: A unit of light flow, or luminous flux. The lumen rating of a lamp is a measure of the total light output of the lamp.

LUMINAIRE: A complete lighting unit consisting of a lamp or lamps, along with the parts designed to distribute the light, hold the lamps, and connect the lamps to a power source. Also called a fixture.

LUMINAIRE EFFICIENCY: The ratio of total lumen output of a luminaire and the lumen output of the lamps, expressed as a percentage. For example, if two luminaires use the same lamps, more light will be emitted from the fixture with the higher efficiency.

LUMINANCE: A photometric term that quantifies brightness of a light source or of an illuminated surface that reflects light. It is expressed as footlamberts (English units) or candelas per square meter (Metric units).

LUX (LX): The metric unit of measure for illuminance of a surface. One lux is equal to one lumen per square meter. One lux equals 0.093 foot-candles.

MAINTAINED ILLUMINANCE: Refers to light levels of a space at other than initial or rated conditions. This term considers light loss factors such as lamp lumen depreciation, luminaire dirt depreciation and room surface dirt depreciation.

MERCURY-VAPOR LAMP: A type of high intensity discharge (HID) lamp in which most of the light is produced by radiation from mercury vapor. Emits a blue-green cast of light. Available in clear and phosphor-coated lamps.

METAL-HALIDE: A type of high intensity discharge (HID) lamp in which most of the light is produced by radiation of metal halide and mercury vapors in the arc tube. Available in clear and phosphor-coated lamps.

LIGHTING TERMS (cont.)

MR-16: A low-voltage quartz reflector lamp, only 2" in diameter. Typically the lamp and reflector are one unit, which directs a sharp, precise beam of light.

NADIR: A reference direction directly below a luminaire, or "straight down" (0 degree angle).

NEMA: Abbreviation for National Electrical Manufacturers Association.

NIST: Abbreviation for National Institute of Standards and Technology.

NPF (NORMAL POWER FACTOR): A ballast/lamp combination in which no components (e.g., capacitors) have been added to correct the power factor, making it normal (essentially low, typically 0.5 or 50%).

OCCUPANCY SENSOR: Control device that turns lights off after the space becomes unoccupied. May be ultrasonic, infrared or other type.

OPTICS: A term referring to the components of a light fixture (such as reflectors, refractors, lenses, louvers) or to the light emitting or light-controlling performance of a fixture.

PAR LAMP: A parabolic aluminized reflector lamp. An incandescent, metal halide, or compact fluorescent lamp used to redirect light from the source using a parabolic reflector. Lamps are available with flood or spot distributions.

PAR 36: A PAR lamp that is 36 one-eighths of an inch in diameter with a parabolic shaped reflector. (SEE **PAR LAMP**)

PARABOLIC LUMINAIRE: A popular type of fluorescent fixture that has a louver composed of aluminum baffles curved in a parabolic shape. The resultant light distribution produced by this shape provides reduced glare, better light control, and is considered to have greater aesthetic appeal.

LIGHTING TERMS (cont.)

PARACUBE: A metallic coated plastic louver made up of small squares. Often used to replace the lens in an installed troffer to enhance its appearance. The paracube is visually comfortable, but the luminaire efficiency is lowered. Also used in rooms with computer screens because of their glare-reducing qualities.

PHOTOCELL: A light sensing device used to control luminaires and dimmers in response to detected light levels.

PHOTOMETRIC REPORT: A photometric report is a set of printed data describing the light distribution, efficiency, and zonal lumen output of a luminaire. This report is generated from laboratory testing.

POWER FACTOR: The ratio of AC volts x amps through a device to AC wattage of the device. A device such as a ballast that measures 120 volts, 1 amp, and 60 watts has a power factor of 50% (volts x amps = 120 VA, therefore 60 watts/120 VA = 0.5). Some utilities charge customers for low power factor systems.

PREHEAT: A type of ballast/lamp circuit that uses a separate starter to heat up a fluorescent lamp before high voltage is applied to start the lamp.

QUAD-TUBE LAMP: A compact fluorescent lamp with a double twin tube configuration.

RADIO FREQUENCY INTERFERENCE (RFI): Interference to the radio frequency band caused by other high frequency equipment or devices in the immediate area. Fluorescent lighting systems generate RFI.

RAPID START (RS): The most popular fluorescent lamp/ballast combination used today. This ballast quickly and efficiently preheats lamp cathodes to start the lamp. Uses a “bi-pin” base.

ROOM CAVITY RATIO (RCR): A ratio of room dimensions used to quantify how light will interact with room surfaces. A factor used in illuminance calculations.

REFLECTANCE: The ratio of light reflected from a surface to the light incident on the surface. Reflectances are often used for lighting calculations. The reflectance of a dark carpet is around 20%, and a clean white wall is roughly 50% to 60%.

REFLECTOR: The part of a light fixture that shrouds the lamps and redirects some light emitted from the lamp.

REFRACTOR: A device used to redirect the light output from a source, primarily by bending the waves of light.

RECESSED: The term used to describe the door frame of a troffer where the lens or louver lies above the surface of the ceiling.

REGULATION: The ability of a ballast to hold constant (or nearly constant) the output watts (light output) during fluctuations in the voltage feeding of the ballast. Normally specified as +/- percent change in output compared to +/- percent change in input.

RELAY: A device that switches an electrical load on or off based on small changes in current or voltage. Examples: low voltage relay and solid state relay.

RETROFIT: Refers to upgrading a fixture, room or building by installing new parts or equipment.

SELF-LUMINOUS EXIT SIGN: An illumination technology using phosphor-coated glass tubes filled with radioactive tritium gas. The exit sign uses no electricity and thus does not need to be hardwired.

SEMI-SPECULAR: Term describing the light reflection characteristics of a material. Some light is reflected directionally, with some amount of scatter.

SHIELDING ANGLE: The angle measured from the ceiling plane to the line of sight where the bare lamp in a luminaire becomes visible. Higher shielding angles reduce direct glare. It is the complementary angle of the cutoff angle. (See **CUTOFF ANGLE**).

LIGHTING TERMS (cont.)

SPACING CRITERION: A maximum distance that interior fixtures may be spaced that ensures uniform illumination on the work plane. The luminaire height above the work plane multiplied by the spacing criterion equals the center-to-center luminaire spacing.

SPECULAR: Mirrored or polished surface. The angle of reflection is equal to the angle of incidence. This word describes the finish of the material used in some louvers and reflectors.

STARTER: A device used with a ballast to start preheat fluorescent lamps.

STROBOSCOPIC EFFECT: Condition where rotating machinery or other rapidly moving objects appear to be standing still due to the alternating current supplied to light sources. Sometimes called "strobe effect."

T12 LAMP: Industry standard for a fluorescent lamp that is 12 one-eighths (1½ inches) in diameter. Other sizes are T10 (1¼ inches) and T8 (1 inch) lamps.

TANDEM WIRING: A wiring option in which a ballast is shared by two or more luminaires. This reduces labor, materials and energy costs. Also called "master-slave" wiring.

THERMAL FACTOR: A factor used in lighting calculations that compensates for the change in light output of a fluorescent lamp due to a change in bulb wall temperature. It is applied when the lamp-ballast combination under consideration is different from that used in the photometric tests.

TRIGGER START: Type of ballast commonly used with 15-watt and 20-watt straight fluorescent lamps.

TROFFER: The term used to refer to a recessed fluorescent light fixture (combination of trough and coffer).

TUNGSTEN HALOGEN LAMP: A gas-filled tungsten filament incandescent lamp with a lamp envelope made of quartz to withstand the high temperature. This lamp contains some halogens (namely iodine, chlorine, bromine and fluorine), which slow the evaporation of the tungsten. Also, commonly called a quartz lamp.

TWIN-TUBE: (SEE **COMPACT FLUORESCENT LAMP**)

ULTRA VIOLET (UV): Invisible radiation that is shorter in wavelength and higher in frequency than visible violet light (literally beyond the violet light).

UNDERWRITERS' LABORATORIES (UL): An independent organization whose responsibilities include rigorous testing of electrical products. When products pass these tests, they can be labeled (and advertised) as "UL listed." UL tests for product safety only.

VANDAL-RESISTANT: Fixtures with rugged housings, break-resistant type shielding, and tamper-proof screws.

VCP: Abbreviation for visual comfort probability. A rating system for evaluating direct discomfort glare. This method is a subjective evaluation of visual comfort expressed as the percent of occupants of a space who will be bothered by direct glare. VCP allows for several factors: luminaire luminances at different angles of view, luminaire size, room size, luminaire mounting height, illuminance, and room surface reflectively. VCP tables are often provided as part of photometric reports.

VERY HIGH OUTPUT (VHO): A fluorescent lamp that operates at a "very high" current (1500 mA), producing more light output than a "high output" lamp (800 mA) or standard output lamp (430 mA).

VOLT: The standard unit of measurement for electrical potential. It defines the "force" or "pressure" of electricity.

VOLTAGE: The difference in electrical potential between two points of an electrical circuit.

LIGHTING TERMS (cont.)

WALLWASHER: Describes luminaires that illuminate vertical surfaces.

WATT (W): The unit for measuring electrical power. It defines the rate of energy consumption by an electrical device when it is in operation. The energy cost of operating an electrical device is calculated as its wattage times the hours of use. In single phase circuits, it is related to volts and amps by the formula: Volts x Amps x PF = Watts. (Note: For AC circuits, PF must be included.)

WORK PLANE: The level at which work is done and at which illuminance is specified and measured. For office applications, this is typically a horizontal plane 30 inches above the floor (desk height).

ZENITH: The direction directly above the luminaire (180° angle).

CHAPTER 7
ELECTRICAL MAINTENANCE

BASIC ELECTRICAL SAFETY GUIDELINES

- Always comply with the NEC®.
- Use UL® approved appliances, components and equipment.
- Keep electrical grounding circuits in good condition. Ground any conductive component or element that does not have to be energized. The grounding connection must be a low-resistance conductor heavy enough to carry the largest fault current that may occur.
- Turn OFF, lock out, and tag disconnect switches when working on any electrical circuit or equipment. Test all circuits after they are turned OFF.
- Use double-insulated power tools or power tools that include a third conductor grounding terminal which provides a path for fault current.
- Always use protective and safety equipment.
- Check conductors, cords, components and equipment for signs of wear or damage.
- Never throw water on an electrical fire. Turn OFF the power and use a Class C rated fire extinguisher.
- Never work alone when working in a dangerous area or with dangerous equipment.
- Learn CPR and first aid.
- Do not work in poorly lighted areas.
- Always use nonconductive ladders. Never use a metal ladder.
- Ensure there are no atmospheric hazards such as flammable dust or vapor in the area.
- Use one hand when working on a live circuit to reduce the chance of an electrical shock passing through the heart and lungs.
- Never bypass or disable fuses or circuit breakers.

OSHA SAFETY COLOR CODES	
Color	**Examples**
Red	Fire protection equipment and apparatus, portable containers of flammable liquids, emergency stop pushbuttons/switches
Yellow	Caution and for marking physical hazards, waste containers for explosive or combustible materials, caution against starting, using, or moving equipment under repair, identification of the starting point or power source of machinery
Orange	Dangerous parts of machines, safety starter buttons, the exposed parts of pulleys, gears, rollers, cutting devices, power jaws
Purple	Radiation hazards
Green	Safety areas and location of first aid equipment

LOCKOUT / TAGOUT GUIDELINES

Electrical power must be turned off when any type of electrical equipment is inspected, serviced, repaired, or replaced. The equipment must also be locked out and tagged out to ensure the safety of personnel working with the equipment.

Lockout is the process of turning off the source of electrical power and installing a lock which prevents the power from being turned ON. Tagout is the process of placing a danger tag on the source of electrical power which indicates that the equipment may not be operated until the danger tag is removed.

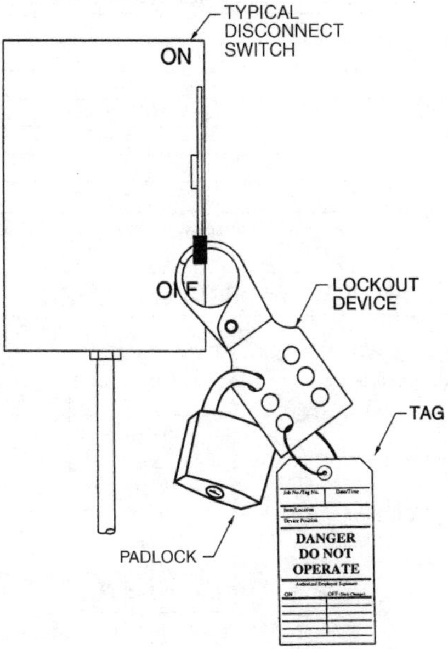

MINIMUM COVER REQUIREMENTS 0-600 VOLTS, NOMINAL

Cover is the distance between the top surface of buried cable, conduit, or raceway and the graded surface.

Minimum Burial	Method of Wiring
18"	Rigid Nonmetallic Conduit (Approved without Concrete Encasement)
6"	Rigid Metal Conduit
6"	Intermediate Metal Conduit
24"	Buried Cables

For details, refer to National Electrical Code®

VOLUME REQUIRED PER CONDUCTOR

Conductor Size (AWG)	Free Space In Box Per Conductor (cu.in.)
18	1.5
16	1.75
14	2
12	2.25
10	2.5
8	3
6	5

Refer to NEC® for complete details.

VERTICAL CONDUCTOR SUPPORTS

Size of Wire (AWG, Circular Mil)	CONDUCTOR TYPE	
	Copper	Aluminum or Copper-Clad Aluminum
18 through 8	100 feet	100 feet
6 through 1/0	100 feet	200 feet
2/0 through 4/0	80 feet	180 feet
Over 4/0 through 350	60 feet	135 feet
Over 350 through 500	50 feet	120 feet
Over 500 through 750	40 feet	95 feet
Over 750	35 feet	85 feet

For SI units: one foot = 0.3048 meter.

MINIMUM DEPTH WORKING CLEARANCES

Nominal Voltage to Ground	Working Conditions 1	2	3
0-150	3'	3'	3'
151-600	3'	3½'	4'
601-2,500	3'	4'	5'
2,501-9,000	4'	5'	6'
9,001-25,000	5'	6'	9'
25,001-75,000	6'	8'	10'
Above 75,000	8'	10'	12'

ELEVATION OF UNGUARDED LIVE PARTS ABOVE WORKING SPACE

Nominal Voltage Between Phases	Elevation
601-7,500	9'
7,501-35,000	9½'
Over 35,000	9½' + .37/inches for every 1,000 volts over 35,000 volts

MINIMUM CLEARANCE – LIVE PARTS

*Minimum Clearance Of Live Parts in Inches				Nominal Voltage Rating in kV	Impulse Withstand B.I.L. in kV	
Phase-To-Phase		Phase-To-Ground				
Indoors	Outdoors	Indoors	Outdoors		Indoors	Outdoors
4.5	7	3.0	6	2.4-4.16	60	95
5.5	7	4.0	6	7.2	75	95
7.5	12	5.0	7	13.8	95	110
9.0	12	6.5	7	14.4	110	110
10.5	15	7.5	10	23	125	150
12.5	15	9.5	10	34.5	150	150
18.0	18	13.0	13		200	200
	18		13	46		200
	21		17			250
	21		17	69		250
	31		25			350
	53		42	115		550
	53		42	138		550
	63		50			650
	63		50	161		650
	72		58			750
	72		58	230		750
	89		71			900
	105		83			1050

The minimum clearance for rigid parts and bare conductors under favorable conditions. They shall be increased for conductor movement or unfavorable conditions, or where space is limited. Impulse withstand voltage for a particular system voltage is determined by the type of surge protection equipment that is utilized.

EFFECT OF ELECTRIC CURRENT ON HUMAN BODY

Current (ma)	Effect	Result
0 to 6	Slight sensation possible	None
6 to 15	Painful shock muscular contraction	Possible no "Let Go"
15 to 20	Painful shock Frozen until circuit is de-energized	No "Let Go"
20 to 50	Severe muscular contractions Asphyxia	Often Fatal
50 to 200	Ventricular fibrillation	Probably Fatal
200+	Heart movement stops	Fatal

1 ma = .001 amp
Duration of current flow is not noted in this comparison

GROUNDING

Electrical circuits are grounded to safeguard equipment and personnel against the hazards of electrical shock. Proper grounding of electrical tools, machines, equipment, and delivery systems is one of the most important factors in preventing hazardous conditions.

GROUNDING ELECTRODE CONDUCTORS – AC SYSTEMS

GROUNDING ELECTRODE CONDUCTOR		SERVICE-ENTRANCE CONDUCTOR OR EQUIVALENT AREA FOR PARALLEL CONDUCTORS	
COPPER	ALUMINUM OR COPPER-CLAD ALUMINUM*	COPPER	ALUMINUM OR COPPER-CLAD ALUMINUM
8 AWG 6 4	6 AWG 4 2	2 OR SMALLER 1 OR 1/0 2/0 OR 3/0 AWG	1/0 OR SMALLER 2/0 OR 3/0 AWG 4/0 OR 250 kcmil
2	1/0	OVER 3/0 THRU 350 kcmil	OVER 250 THRU 500 kcmil
1/0	3/0	OVER 350 kcmil THRU 600 kcmil	OVER 500 kcmil THRU 900 kcmil
2/0	4/0	OVER 600 kcmil THRU 1100 kcmil	OVER 900 kcmil THRU 1750 kcmil
3/0	250 kcmil	OVER 1100 kcmil	OVER 1750 kcmil

A) The table above applies to the derived conductors of separately derived AC systems.

B) When multiple sets of service conductors are utilized, the equivalent size of the largest service-entrance conductor shall be determined by the largest sum of the areas of the corresponding conductors of each set.

C) If there are no service-entrance conductors, the grounding electrode conductor size shall be determined by the equivalent size of the largest service-entrance conductor required for the load to be served.

*NOTE: Please refer to NEC® Installation restrictions concerning aluminum and copper-clad aluminum conductors.

MINIMUM SIZE CONDUCTORS – GROUNDING RACEWAY AND EQUIPMENT

CONDUCTOR SIZE		AMPERAGE RATING OR SETTING OF AUTOMATIC OVERCURRENT DEVICE NOT TO EXCEED
COPPER	ALUMINUM OR COPPER-CLAD ALUMINUM	
14 AWG	12 AWG	15 amps
12	10	20 amps
10	8	30 amps
10	8	40 amps
10	8	60 amps
8	6	100 amps
6	4	200 amps
4	2	300 amps
3	1	400 amps
2	1/0	500 amps
1	2/0	600 amps
1/0	3/0	800 amps
2/0	4/0	1000 amps
3/0	250 kcmil	1200 amps
4/0	350	1600 amps
250 kcmil	400	2000 amps
350	600	2500 amps
400	600	3000 amps
500	800	4000 amps
700	1200	5000 amps
800	1200	6000 amps

The equipment grounding conductor shall be sized larger than this table per NEC® installation restrictions.

TYPES OF GROUNDING METHODS PER THE NEC

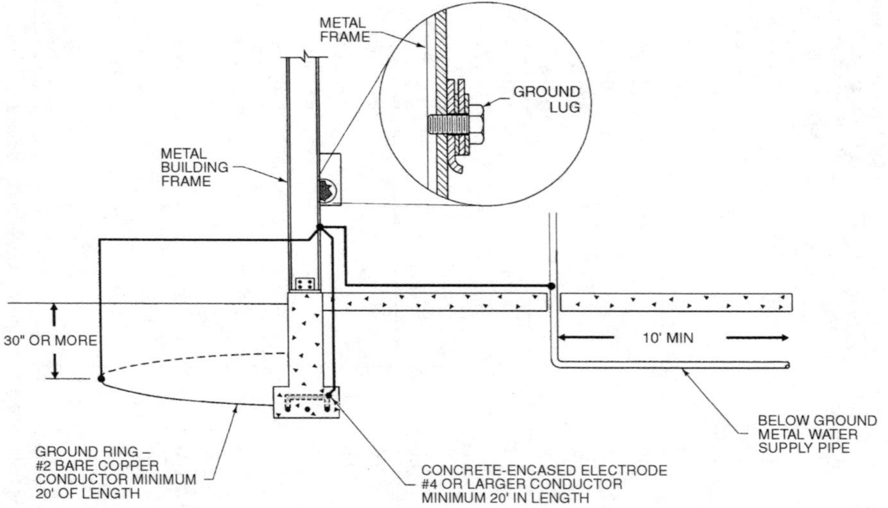

GROUNDING DIFFERENT TYPES OF CIRCUITS

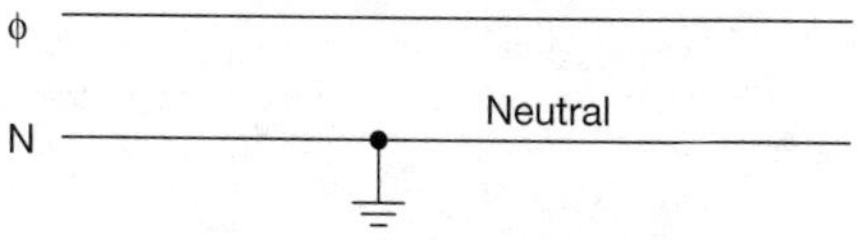

Single-Phase, Two-Wire

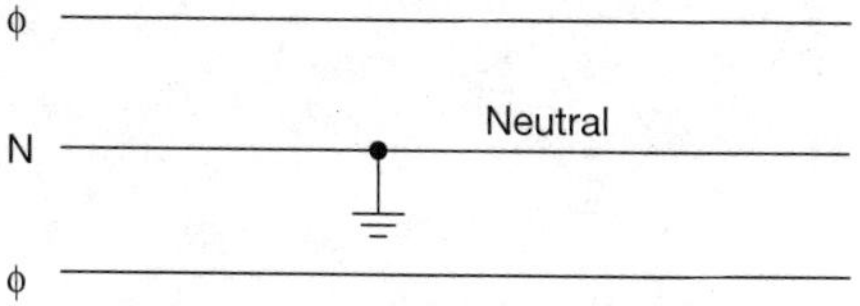

Single-Phase, Three-Wire

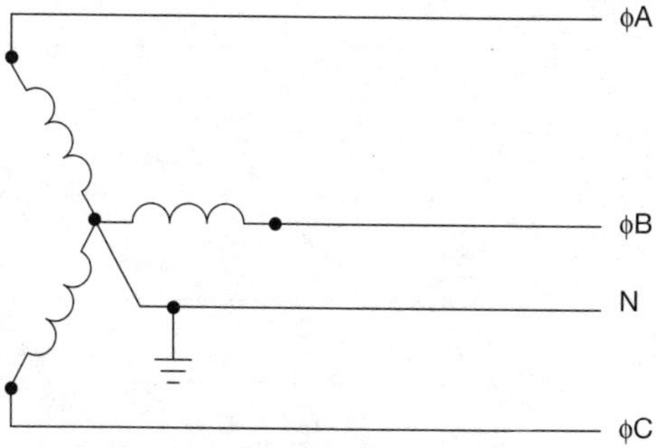

Three-Phase Wye

GROUNDING DIFFERENT TYPES OF CIRCUITS (cont.)

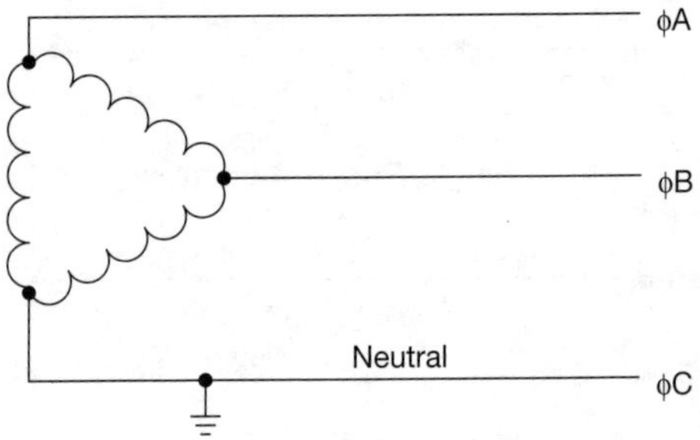

Three-Phase Delta, Three-Wire

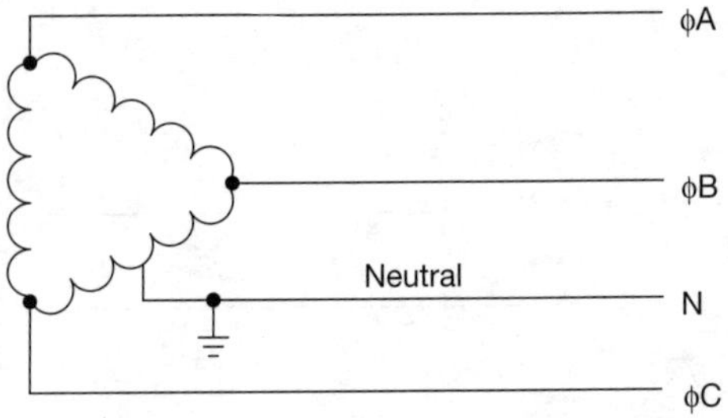

Three-Phase Delta, Four-Wire

VOLTAGE RELATIONSHIP ON GROUNDED 4-WIRE SYSTEMS

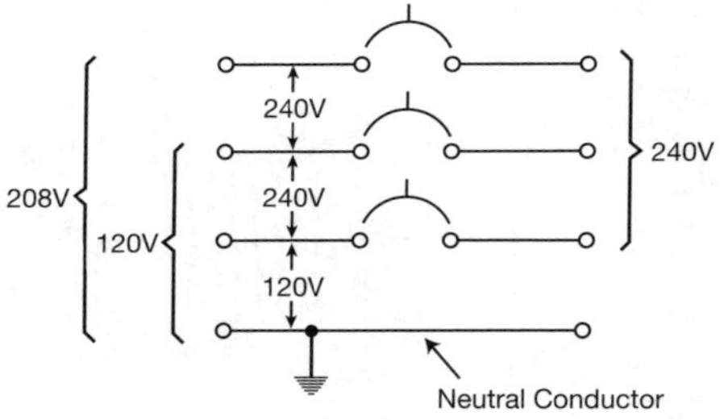

4-wire system with a neutral

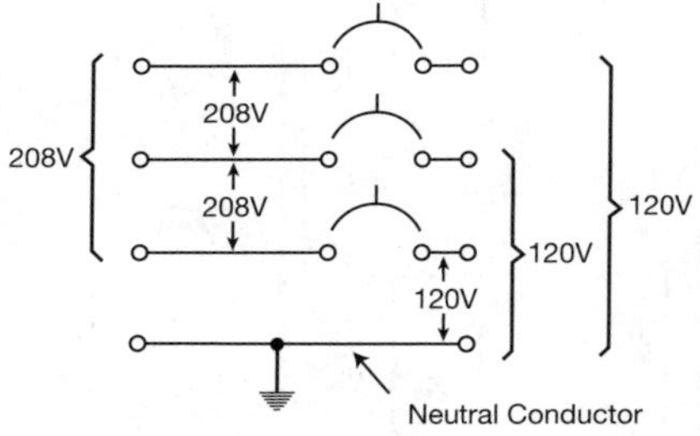

Grounded 4-wire wye system

GROUNDING AN EXISTING CIRCUIT

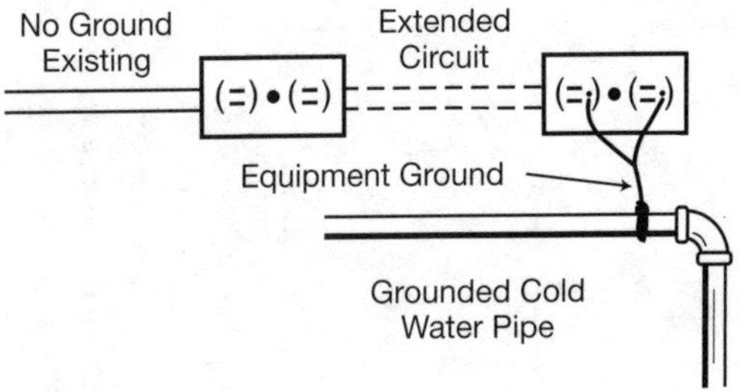

GROUNDING A SCREW-SHELLBASE

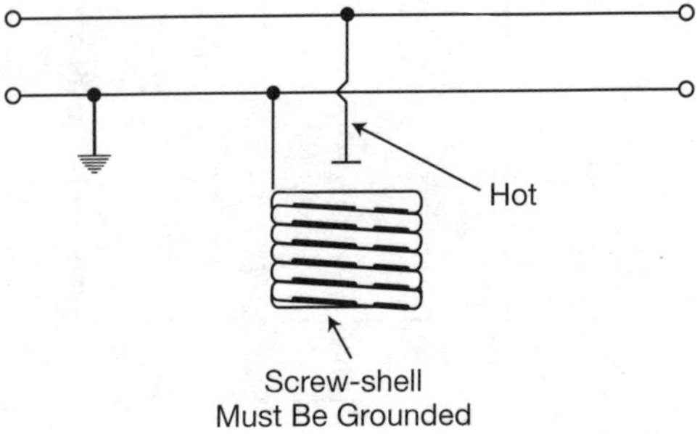

DECIBEL LEVELS OF SOUNDS

The definition of sound intensity is energy (erg) transmitted per 1 second over a square centimeter surface. Sounds are measured in decibels. A decibel (db) change of 1 is the smallest change detected by humans.

Hearing Intensity	Decibel Level	Examples of Sounds
Barely	0	Dead silence
Audible		Audible hearing threshold
	10	Room (sound proof)
(Very	20	Empty auditorium
Light)		Ticking of a stopwatch
		Soft whispering
Audible	30	People talking quietly
Light	40	Quiet street noise without autos
Medium	45	Telephone operator
Loud	50	Fax machine in office
	60	Close conversation
Loud	70	Stereo system
		Computer printer
	80	Fire truck / Ambulance siren
		Cat / dog fight
Extremely	90	Industrial machinery
Loud		High school marching band
Damage	100	Heavy duty grinder in a
Possible		machine / welding shop
Damaging	100+	Begins ear damage
	110	Diesel engine of a train
	120	Lightning strike (thunderstorm)
		60 ton metal forming factory press
	130	60" fan in a bus vacuum system
	140	Commercial / Military jet engine
Ear Drum	194	Space shuttle engines
Shattering	225	16" Guns on a battleship

MAXIMUM AVERAGE SOUND LEVELS FOR TRANSFORMERS IN DECIBELS

kVA	Dry-Type		Liquid-Filled	
	Self-Cooled Rating (AA)	Forced-Air Cooling (FA)	Self-Cooled Rating (OA)	Forced-Air Cooling (FA)
0-50	50	–	–	–
51-150	55	–	–	–
151-300	58	67	55	67
301-500	60	67	56	67
501-700	62	67	57	67
701-1000	64	67	58	67
1001-1500	65	68	60	67
1501-2000	66	69	61	67
2001-2500	68	71	62	67
2501-3000	70	71	63	67
3001-4000	71	73	64	67
4001-5000	72	74	65	67
5001-6000	73	75	66	68
6001-7500	–	76	67	69
7501-10000	–	76	68	70

HEARING PROTECTION LEVELS

Hearing protection is mandatory if the following time exposures to decibel levels are exceeded because of possible damage to human hearing.

Decibel Level	Time Exposure Per Day
115	15 minutes
110	30 minutes
105	1 hour
102	1½ hours
100	2 hours
97	3 hours
95	4 hours
92	6 hours
90	8 hours

TYPES OF FIRE EXTINGUISHERS

Today they are virtually standard equipment in a business or residence and are rated by the make up of the fire they will extinguish.

TYPE A: To extinguish fires involving trash, cloth, paper and other wood or pulp based materials. The flames are put out by water based ingredients or dry chemicals.

TYPE B: To extinguish fires involving greases, paints, solvents, gas and other petroleum based liquids. The flames are put out by cutting off oxygen and stopping the release of flammable vapors. Dry chemicals, foams and halon are used.

TYPE C: To extinguish fires involving electricity. The combustion is put out the same way as with a type B extinguisher, but, most importantly, the chemical in a type C MUST be non-conductive to electricity in order to be safe and effective.

TYPE D: To extinguish fires involving combustible metals. Please be advised to obtain important information from your local fire department on the requirements for type D fire extinguishers for your area.

Any combination of letters indicate that an extinguisher will put out more than one type of fire. A type BC will put out two types of fires. The size of the fire to be extinguished is shown by a number in front of the letter such as 100A.
The following formulas apply:

Class "1A": Will extinguish 25 burning sticks 40 inches long.

Class "1B": Will extinguish a paint thinner fire 2.5 square feet in size.

A 100B fire extinguisher will put out a fire 100 times larger than a type 1B.

Here are some basic guidelines to follow:

- By using a type ABC you will cover most basic fires.
- Use fire extinguishers with a gauge and ones that are constructed with metal. Also note if the unit is U.L. approved.
- Utilize more than one extinguisher and be sure that each unit is mounted in a clearly visible and accessible manner.
- After purchasing any fire extinguisher always review the basic instructions for its intended use. Never deviate from the manufacturers' guidelines. Following this simple procedure could end up saving lives.

ELECTRICAL CABLE CLASS RATINGS

Electrical cable is rated according to the following parameters. The number of wires, the wire size, the type of insulation and the moisture condition of the environment of the wire. Therefore, an electrical cable designated 10/2 with ground - type UF - 600volt-(UL) meets the specifications below:

- The "10" relates to wire size - 10 gauge wire.
- 2 defines an electrical cable with two wires.
- The word "Ground" indicates the cable has a third wire to be connected to ground.
- The term "Type UF" means the insulation type has an acceptable moisture rating.
- "600 V" defines the cable as being rated at 600 volts maximum.
- ("UL") means the cable has certification from Underwriters Laboratory.

CABLE INSULATION MOISTURE RATINGS

DRY – Indoor above ground level; moisture usually not encountered.

MOIST – Indoor below ground level (Basement); locations are partially protected; moisture level is moderate.

WET – Locations affected by weather (Outside); concrete slabs, underground, etc.; Water saturation likely.

CONDUCTOR PREFIX CODES

B – Outer braid.

F – Fixture wire.

FEP – Fluorinated ethylene propylene. Use in dry locations only, over 90° C.

H – Load temp up to 75° C.

HH – Load temp up to 90° C.

L – Seamless lead jacket.

M – Machine tool wire.

N – Resistant to oil and gas.

O – Neoprene jacket.

R – Rubber covering.

S – Appliance cord.

SP – Lamp cord, rubber.

SPT – Lamp cord, plastic.

T – Load temp up to 60° C.

W – Wet use only.

X – Moisture and heat resistant.

TYPES OF CONDUCTORS

TYPE	MAX. TEMP	APPLICATION	INSULATION	OUTER COVERING
FEP or FEPB	90°C (194°F)	Dry and damp locations	Fluorinated ethylene propylene	None or Glass braid
	200°C (392°F)	Dry locations – Special Apps		
MI	90°C (194°F)	Dry and wet locations	Magnesium oxide	Copper or alloy steel
	250°C (482°F)	Special Apps		
MTW	60°C (140°F)	Machine tool wiring – wet locations	Flame-retardant, moisture, heat, and oil-resistant thermoplastic	None or Nylon jacket
	90°C (194°F)	Machine tool wiring – dry locations		
PAPER	85°C (185°F)	Underground service conductors	Paper	Lead sheath
PFA	90°C (194°F)	Dry and damp locations	Perfluoroalkoxy	None
	200°C (392°F)	Dry locations – Special Apps		
PFAH	250°C (482°F)	Dry locations only	Perfluoroalkoxy	None
RHH	90°C (194°F)	Dry and damp locations		Moisture resistant, flame-retardant non-metallic
RHW	75°C (167°F)	Dry and wet locations	Flame-retardant, moisture-resistant thermoset	Moisture-resistant, flame-retardant, non-metallic

TYPES OF CONDUCTORS (cont.)

TYPE	MAX. TEMP	APPLICATION	INSULATION	OUTER COVERING
RHW-2	90°C (194°F)	Dry and wet locations	Flame-retardant, moisture-resistant thermoset	Moisture-resistant, flame-retardant, non-metallic
SA	90°C (194°F) 200°C (392°F)	Dry and damp locations Special Apps	Silicone rubber	Glass or braid material
SIS	90°C (194°F)	Switchboard wiring	Flame-retardant thermostat	None
TBS	90°C (194°F)	Switchboard wiring	Thermoplastic	Flame-retardant, non-metallic
TFE	250°C (482°F)	Dry locations only	Extruded polytetrafluoroethylene	None
THHN	90°C (194°F)	Dry and damp locations	Flame-retardant, heat resistant thermoplastic	Nylon jacket
THHW	75°C (167°F) 90°C (194°F)	Wet locations Dry locations	Flame-retardant, moisture- and heat-resistant thermoplastic	None
THW	75°C (167°F) 90°C (194°F)	Dry and wet locations Special Apps	Flame-retardant, moisture- and heat-resistant thermoplastic	None
THWN	75°C (167°F)	Dry and wet locations	Flame-retardant, moisture- and heat-resistant thermoplastic	Nylon jacket
TW	60°C (140°F)	Dry and wet locations	Flame-retardant, moisture- and heat-resistant thermoplastic	None
UF	60°C (140°F) 75°C (167°F)	Refer to NEC®	Moisture-resistant Moisture and heat resistant	Integral with insulation

TYPES OF CONDUCTORS (cont.)

TYPE	MAX. TEMP	APPLICATION	INSULATION	OUTER COVERING
USE	75°C (167°F)	Refer to NEC®	Heat and moisture-resistant	Moisture-resistant non-metallic
XHH	90°C (194°F)	Dry and damp locations	Flame-retardant thermoplastic	None
XHHW	90°C (194°F) 75°C (167°F)	Dry and damp locations Wet locations	Flame-retardant, moisture-resistant thermoset	None
XHHW-2	90°C (194°F)	Dry and wet locations	Flame-retardant, moisture-resistant thermoset	None
Z	90°C (194°F) 150°C (302°F)	Dry and damp locations Dry locations – Special Apps	Modified ethylene tetrafluoro ethylene	None
ZW	75°C (167°F) 90°C (194°F) 150°C (302°F)	Wet locations Dry and damp locations Dry locations – Special Apps	Modified ethylene tetrafluoro ethylene	None

CABLE JACKET MATERIALS

Mechanical	PVC	Polyethylene	Neoprene	Chlorosulphonated Polyethylene	Thermoplastic CPE
Abrasion Resistance	Good	Excellent	Good	Good	Excellent
Tensile Strength	Excellent	Excellent	Excellent	Excellent	Good
Elongation	Good	Excellent	Excellent	Excellent	Good
Compression Resistance	Good	Excellent	Excellent	Excellent	Good
Flexibility	Good	Fair	Excellent	Excellent	Fair
Environmental					
Flame	Good	Poor	Excellent	Excellent	Good
Moisture Fresh or salt water	Good	Exceptional	Good	Excellent	Excellent
Petroleum oils Motor oil Fuel oil Crude oil	Good	Excellent (Slight swelling above 60°C)	Good	Good	Good (Poor above 110°C)
Creosote	Poor	Good	Fair	Fair	Good
Paraffinic Hydrocarbons Gasoline Kerosene	Good	Excellent (Slight swelling at higher temperatures)	Poor	Poor	Excellent (Slight swelling at higher temperatures)

CABLE JACKET MATERIALS (cont.)

Environmental	PVC	Polyethylene	Neoprene	Chlorosulphonated Polyethylene	Thermoplastic CPE
Alcohols Isopropyl Wood Grain	Fair	Good	Fair	Good	Good
Mineral Acids Sulfuric Nitric Hydrochloric	Excellent	Excellent	Excellent	Excellent	Excellent
Fixed Alkalies Sodium hydroxide (lye) Potassium hydroxide (potash) Calcium hydroxide (lime)	Good	Excellent	Good	Excellent	Excellent
Ketones Acetone Methyl ethyl ketone (MEK)	Poor	Good	Poor	Fair	Good
Esters Ethyl Acetate Most lacquer thinners	Poor	Good	Poor	Fair	Good
Halogenated Hydrocarbons Chloroform Carbon tetrachloride Methyl chloride	Poor	Poor	Poor	Poor	Poor

CABLE JACKET MATERIALS (cont.)

General	PVC	Polyethylene	Neoprene	Chlorosulphonated Polyethylene	Thermoplastic CPE
Leaves protective residue after combustion	Yes	No	Yes	Yes	Yes
Oxygen Index (ASTM D-2863)	23-30%	17-18%	31-39%	30-36%	30-34%
Halogen content – % Wt.	26	0	18	14	18-20
Minimum installation temperature	14°F (-10°C)	-40°F (-40°C)	-4°F (-20°C)	-4°F (-20°C)	-40°F (-40°C)
Dimensional stability under heat	Fair	Fair	Excellent	Excellent	Fair
Maximum operating temperature	75°C (167°F)	75°C (167°F)	90°C (194°F)	90°C (194°F)	75°C (167°F)

SIZE OF EXTENSION CORDS FOR PORTABLE ELECTRIC TOOLS FOR 115-VOLT

Full-Load Ampere Rating of Tool	0 to 3.4 A	3.5 to 5 A	5.1 to 7 A	7.1 to 12 A	12.1 to 16 A
Length of Cord	**Wire Size (AWG)**				
25 feet	18	18	16	14	14
50 feet	18	18	16	14	12
75 feet	18	16	14	12	10
100 feet	16	14	12	10	8
200 feet	14	12	10	8	6
300 feet	12	10	8	6	4
400 feet	10	8	6	4	4
500 feet	10	8	6	4	2
600 feet	8	6	4	2	2
800 feet	8	6	4	2	1
1000 feet	6	4	2	1	0

***NOTE:** If voltage is already low at the outlet (source), increase to standard voltage or use a larger cable than shown.*

AMAPACITY OF LAMP AND EXTENSION CORDS (A.W.G.) - TYPES S, SJ, SJT, SP, SPT, ST

Gauge	4 Conductors	3 Conductors	2 Conductors
18	6	7	10
16	8	10	13
14	12	15	18
12	16	20	25
10	20	25	30

ENCLOSURES

Type	Service Conditions	Sealing Method	Cost
1	No unusual		Base
4	Windblown dust and rain, splashing water, hose-directed water, and ice on enclosure		12 x Base
4X	Corrosion, windblown dust and rain, splashing water, hose-directed water, and ice on enclosure.		12 x Base
7	Withstand and contain an internal explosion of specified gases, contain an explosion sufficiently so an explosive gas-air mixture in the atmosphere is not ignited.		48 x Base
9	Dust		48 x Base
12	Dust, falling dirt, and dripping noncorrosive liquids		5 x Base

HAZARDOUS LOCATIONS

Class	Group	Material
I	A	Acetylene
	B	Hydrogen, butadiene, ethylene oxide, propylene oxide
	C	Carbon monoxide, ether, ethylene, hydrogen sulfide, morpholine, cyclopropane
	D	Gasoline, benzene, butane, propane, alcohol, acetone, ammonia, vinyl chloride
II	E	Metal dusts
	F	Carbon black, coke dust, coal
	G	Grain dust, flour, starch, sugar, plastics
III	No groups	Wood chips, cotton, flax, and nylon

ENCLOSURE TYPES

Type	Use	Service Conditions	UL Tests	Comments
1	Indoor	None	Rod entry, rust resistance	
3	Outdoor	Windblown dust, rain, sleet, and ice on enclosure	Rain, external icing, dust, and rust resistance	Do not provide protection against internal condensation, or internal icing
3R	Outdoor	Falling rain and ice on enclosure	Rod entry, rain, external icing, and rust resistance	Do not provide protection against dust, internal condensation, or internal icing
4	Indoor/outdoor	Windblown dust and rain, splashing water, hose-directed water and ice on enclosure	Hosedown, external icing and rust resistance	Do not provide protection against internal condensation, or internal icing
4X	Indoor/outdoor	Corrosion, windblown dust and rain, splashing water, hose-directed water and ice on enclosure	Hosedown, external icing, and corrosion resistance	Do not provide protection against internal condensation, or internal icing
6	Indoor/outdoor	Occasional temporary submersion at a limited depth		
6P	Indoor/outdoor	Prolonged submersion at a limited depth		

ENCLOSURE TYPES (cont.)

Type	Use	Service Conditions	UL Tests	Comments
7	Indoor locations classified as Class I, or Groups A, B, C, or D, as defined in the NEC®	Withstand and contain an internal explosion of specified gases, contain an explosion sufficiently so an explosive gas-air mixture in the atmosphere is not ignited	Explosion, hydrostatic, and temperature	Enclosed heat-generating devices shall not cause external surfaces to reach temperatures capable of igniting explosive gas-air mixtures in the atmosphere
9	Indoor locations classified as Class II, Groups E or G, as defined in the NEC®	Dust	Dust penetration, temperature, and gasket aging	Enclosed heat-generating devices shall not cause external surfaces to reach temperatures capable of igniting explosive gas-air mixtures in the atmosphere
12	Indoor	Dust, falling dirt, and dripping noncorrosive liquids	Drip, dust, and rust resistance	Do not provide protection against internal condensation
13	Indoor	Dust, spraying water, oil and noncorrosive coolant	Oil explosion and rust resistance	Do not provide protection against internal condensation

NON-LOCKING WIRING DEVICES

2-POLE, 3-WIRE

RECEPTACLE CONFIGURATION	WIRING DIAGRAM	RATING	NEMA ANSI
	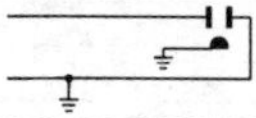	15 A 125 V	5-15 C73.11
	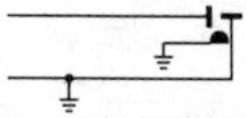	20 A 125 V	5-20 C73.12
	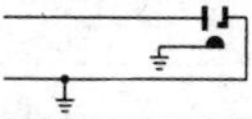	30 A 125 V	5-30 C73.45
	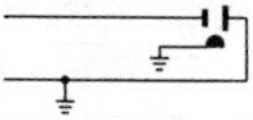	50 A 125 V	5-50 C73.46
	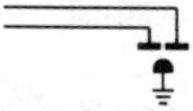	15 A 250 V	6-15 C73.20
	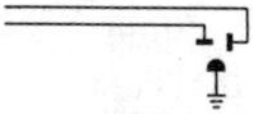	20 A 250 V	6-20 C73.51
	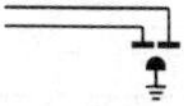	30 A 250 V	6-30 C73.52
	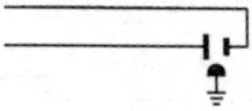	50 A 250 V	6-50 C73.53
	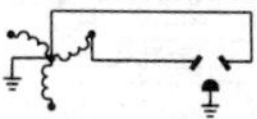	15 A 277 V	7-15 C73.28
	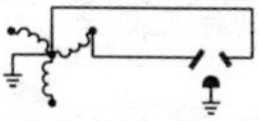	20 A 277 V	7-20 C73.63
	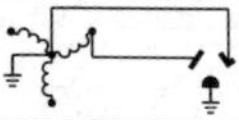	30 A 277 V	7-30 C73.64
	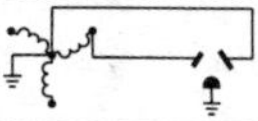	50 A 277 V	7-50 C73.65

NON-LOCKING WIRING DEVICES (cont.)

4-POLE, 4-WIRE

RECEPTACLE CONFIGURATION	WIRING DIAGRAM	RATING	NEMA ANSI
	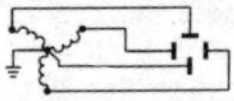	15 A 3 φ Y 120/208 V	18-15 C73.15
	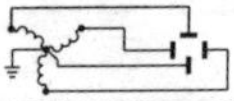	20 A 3 φ Y 120/208 V	18-20 C73.26
	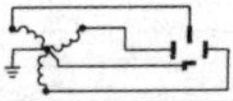	30 A 3 φ Y 120/208 V	18-30 C73.47
	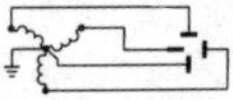	50 A 3 φ Y 120/208 V	18-50 C73.48
	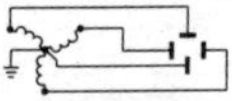	60 A 3 φ Y 120/208 V	18-60 C73.27

3-POLE, 3-WIRE

RECEPTACLE CONFIGURATION	WIRING DIAGRAM	RATING	NEMA ANSI
	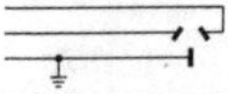	20 A 125/250 V	10-20 C73.23
	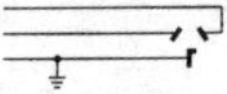	30 A 125/250 V	10-30 C73.24
	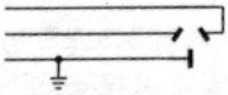	50 A 125/250 V	10-50 C73.25
	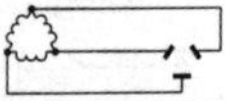	15 A 3φ 250 V	11-15 C73.54
	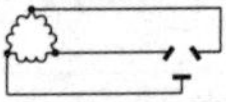	20 A 3φ 250 V	11-20 C73.55
	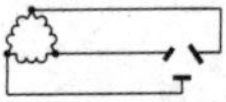	30 A 3φ 250 V	11-30 C73.56
	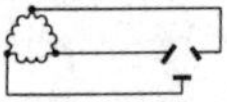	50 A 3φ 250 V	11-50 C73.57

NON-LOCKING WIRING DEVICES (cont.)

3-POLE, 4-WIRE

RECEPTACLE CONFIGURATION	WIRING DIAGRAM	RATING	NEMA ANSI
	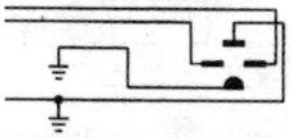	15 A 125/250 V	14-15 C73.49
	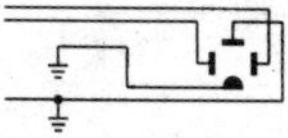	20 A 125/250 V	14-20 C73.50
	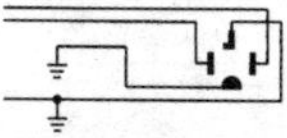	30 A 125/250 V	14-30 C73.16
	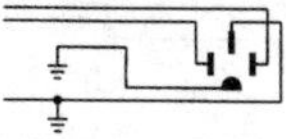	50 A 125/250 V	14-50 C73.17
	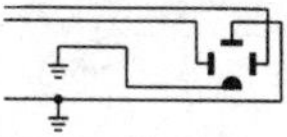	60 A 125/250 V	14-60 C73.18
	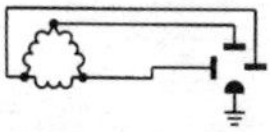	15 A 3φ 250 V	15-15 C73.58
	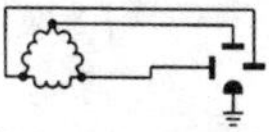	20 A 3φ 250 V	15-20 C73.59
	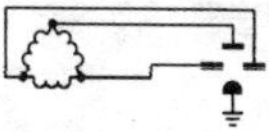	30 A 3φ 250 V	15-30 C73.60
	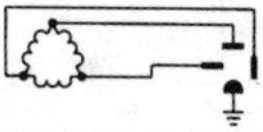	50 A 3φ 250 V	15-50 C73.61
	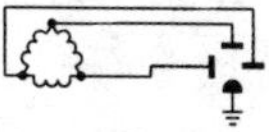	60 A 3φ 250 V	15-60 C73.62

LOCKING WIRING DEVICES

2-POLE, 3-WIRE

RECEPTACLE CONFIGURATION	WIRING DIAGRAM	RATING	NEMA ANSI
	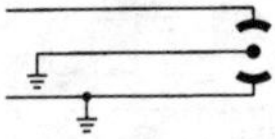	15 A 125 V	ML2 C73.44
	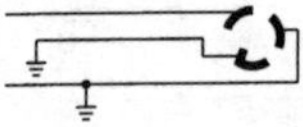	15 A 125 V	L5-15 C73.42
	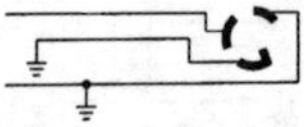	20 A 125 V	L5-20 C73.72
	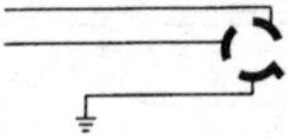	15 A 250 V	L6-15 C73.74
	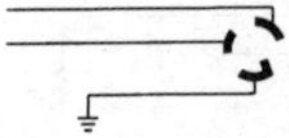	20 A 250 V	L6-20 C73.75
	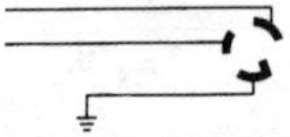	30 A 250 V	L6-30 C73.76
	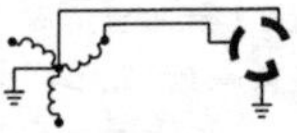	15 A 277 V	L7-15 C73.43
	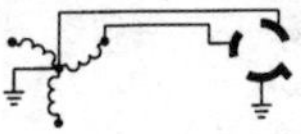	20 A 277 V	L7-20 C73.77
	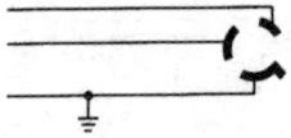	20 A 480 V	L8-20 C73.79
	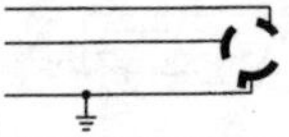	20 A 600 V	L9-20 C73.81

LOCKING WIRING DEVICES (cont.)

3-POLE, 4-WIRE

RECEPTACLE CONFIGURATION	WIRING DIAGRAM	RATING	NEMA ANSI
	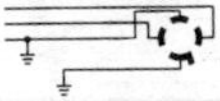	20 A 125/250 V	L14-20 C73.83
	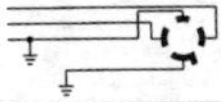	30 A 125/250 V	L14-30 C73.84
	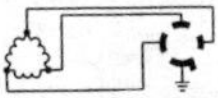	20 A 3ϕ 250 V	L15-20 C73.85
	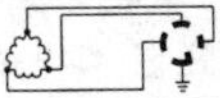	30 A 3ϕ 250 V	L15-30 C73.86
	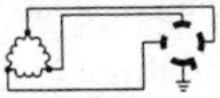	20 A 3ϕ 480 V	L16-20 C73.87
	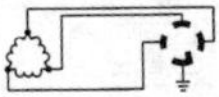	30 A 3ϕ 480 V	L16-30 C73.88
	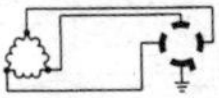	30 A 3ϕ 600 V	L17-30 C73.89

3-POLE, 3-WIRE

RECEPTACLE CONFIGURATION	WIRING DIAGRAM	RATING	NEMA ANSI
	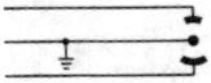	15 A 125/250 V	ML3 C73.30
	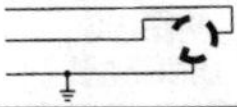	20 A 125/250 V	L10-20 C73.96
	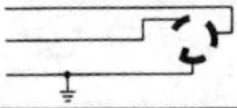	30 A 125/250 V	L10-30 C73.97
	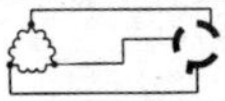	15 A 3ϕ 250 V	L11-15 C73.98
	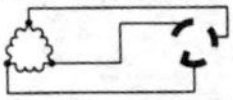	20 A 3ϕ 250 V	L11-20 C73.99

LOCKING WIRING DEVICES (cont.)

3-POLE, 3-WIRE (cont.)

RECEPTACLE CONFIGURATION	WIRING DIAGRAM	RATING	NEMA ANSI
	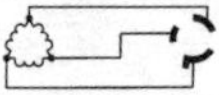	20 A 3ϕ 480 V	L12-20 C73.101
	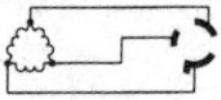	30 A 3ϕ 480 V	L12-30 C73.102
	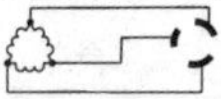	30 A 3ϕ 600 V	L13-30 C73.103

4-POLE, 4-WIRE

RECEPTACLE CONFIGURATION	WIRING DIAGRAM	RATING	NEMA ANSI
	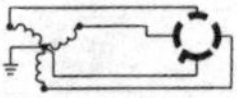	20 A 3 ϕ Y 120/208 V	L18-20 C73.104
	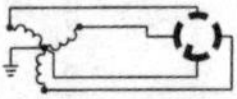	30 A 3 ϕ Y 120/208 V	L18-30 C73.105
	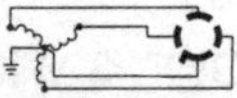	20 A 3 ϕ Y 277/480 V	L19-20 C73.106
	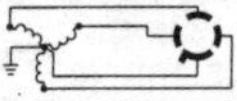	20 A 3 ϕ Y 347/600 V	L20-20 C73.108

4-POLE, 5-WIRE

RECEPTACLE CONFIGURATION	WIRING DIAGRAM	RATING	NEMA ANSI
	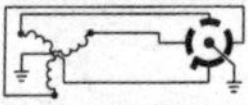	20 A 3 ϕ Y 120/208 V	L21-20 C73.90
	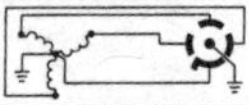	20 A 3 ϕ Y 277/480 V	L22-20 C73.92
	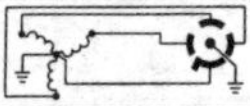	20 A 3 ϕ Y 347/600 V	L23-20 C73.94

SIZING SINGLE-PHASE TRANSFORMERS

1. Determine the total voltage required by the loads.
2. Determine the kVA capacity required by the loads.
3. Check frequency of the supply voltage and the loads.
4. Check the supply voltage with the rating of the primary side.

To calculate kVA capacity of a 1ϕ transformer

$$kVA_{CAP} = E \times \frac{I}{1000}$$

WHERE: kVA_{CAP} = transformer capacity (in kVA)

E = voltage (in Volts)

I = current (in Amps)

The transformer must have a kVA capacity 10% greater than that required by the loads.

1ϕ FULL LOAD CURRENTS (AMPS)						
kVA	**120 V**	**208 V**	**240 V**	**277 V**	**380 V**	**480 V**
.050	.4	.2	.2	.2	.1	.1
.100	.8	.5	.4	.3	.2	.2
.150	1.2	.7	.6	.5	.4	.3
.250	2.0	1.2	1	.9	.6	.5
.500	4.2	2.4	2.1	1.8	1.3	1
.750	6.3	3.6	3.1	2.7	2	1.6
1	8.3	4.8	4.2	3.6	2.6	2.1
1.5	12.5	7.2	6.2	5.4	3.9	3.1
2	16.7	9.6	8.3	7.2	5.2	4.2
3	25	14.4	12.5	10.8	7.9	6.2
5	41	24	20.8	18	13.1	10.4
7.5	62	36	31	27	19.7	15.6
10	83	48	41	36	26	20.8
15	125	72	62	54	39	31

SIZING THREE-PHASE TRANSFORMERS

To calculate kVA capacity of a 3ϕ transformer

$$kVA_{CAP} = E \text{ x } 1.732 \text{ x } \frac{I}{1000}$$

WHERE: kVA_{CAP} = transformer capacity (in kVA)
E = voltage (in Volts)
1.732 = constant (for 3ϕ power)
I = current (in Amps)

3ϕ FULL LOAD CURRENTS (AMPS)				
kVA	**208 V**	**240 V**	**480 V**	**600 V**
3	8.3	7.2	3.6	2.9
4	12.5	10.8	5.4	4.3
6	16.6	14.4	7.2	5.8
9	25	21.6	10.8	8.6
15	41	36	18	14.4
22	62	54	27	21.6
30	83	72	36	28
45	124	108	54	43
50	139	120	60.2	48.2
60	167	145	72.3	57.8
75	208	180	90	72
100	278	241	120	96.3
112.5	312	270	135	108
150	415	360	180	144
200	554	480	240	192
225	625	540	270	216
300	830	720	360	288
400	1110	960	480	384
500	1380	1200	600	480
750	2080	1800	900	720
1000	2780	2400	1200	960
1500	4150	3600	1800	1440
2000	5540	4800	2400	1920

TRANSFORMER DERATINGS

Maximum Ambient Temperature (°C)	Maximum Transformer Loading (%)
40	100
45	96
50	92
55	88
60	81
65	80
70	76

DERATED KVA CAPACITY

To calculate the derated kVA capacity of a transformer operating at a higher-than-normal ambient temperature condition, use the formula below:

$$kVA = rated\ kVA \times maximum\ load$$

WHERE: *kVA* = derated transformer capacity (in kVA)

rated kVA = manufacturer transformer rating (in kVA)

maximum load = maximum transformer loading (in %)

TRANSFORMER OVERLOADING

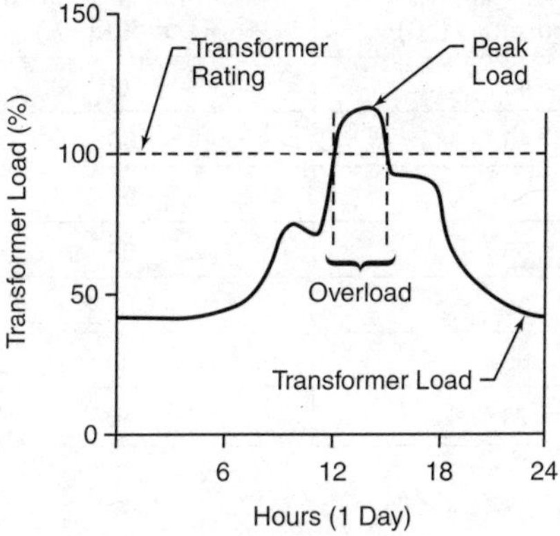

PERMISSIBLE OVERLOADING

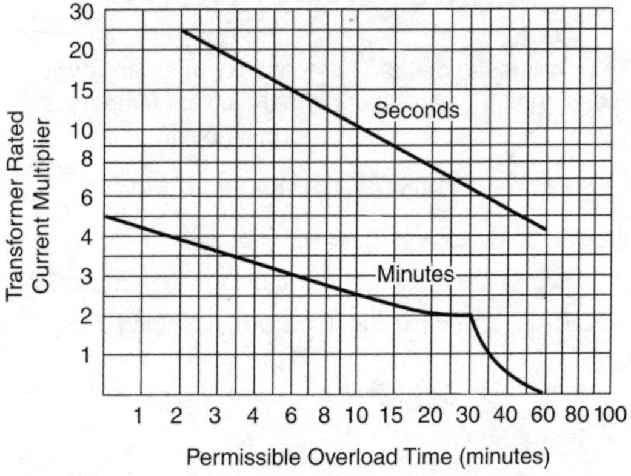

TRANSFORMER AND REGULATOR INSPECTION CHECKLIST

Item	Inspection Interval							
	Large Units				Small Units			
	Attended		Unattended		Attended		Unattended	
Foundation, Rails, and Trucks		A		A		A		A
Tanks and Radiators	D	A	W	A	W	A	M	A
Oil and Water Piping	D	A	W	A	W	A	M	A
Valves and Plugs	D	A	W	A	W	A	M	A
Oil Level, Gauges, and Relays	D	A	W	A	W	A	M	A
Breathers and Vents	D	A	W	A	W	A	M	A
Relief Diaphragm	D	A	W	A	W	A	M	A
Water-Cooling Coils and Piping		A		A		A		A
Flow Indicators and Relays	D	A		A	D	A		A
Heat Exchangers		A						
Oil Pumps	D	A						
Cooling Fans	D	A			D	A		

D – routine daily inspection
W – routine weekly inspection
M – routine monthly inspection
Q – quarterly inspection
A – annual inspection
NS – not scheduled

TRANSFORMER AND REGULATOR INSPECTION CHECKLIST (cont.)

Item	Inspection Interval							
	Large Units				Small Units			
	Attended		Unattended		Attended		Unattended	
Temperature Indicators and Relays	D	A	W	A	W	A	M	A
Inert Gas Tanks	D		W					
Gas Regulator, Gauges, and Relays	D	A	W	A				
Gas Piping and Valves		A		A				
Gas Analysis		Q		Q				
Bushings	W	A	W	A	W	A	M	D
Bushing Current Transformers and Potential Device		A		A				
Main Terminal and Ground Connections	D	A	W	A	W	A	M	A
Core and Coils		NS		NS		NS		NS
Internal Inspection		A		A		A		A
Terminal Board and Connections		A		A		A		A
Ratio Adjuster	W	A	W	A	W	A	M	A

D – routine daily inspection
W – routine weekly inspection
M – routine monthly inspection
Q – quarterly inspection
A – annual inspection
NS – not scheduled

TRANSFORMER AND REGULATOR INSPECTION CHECKLIST (cont.)

Item	Inspection Interval							
	Large Units				Small Units			
	Attended		Unattended		Attended		Unattended	
Tap Changer or Regulator	D	A	W	A				
Motor and Drive		A		A				
Auxiliary and Limit Switches		A		A				
Position Indicators		A		A				
Operation Counter	W	A	W	A				
Operation		A		A				
Power Supplies and Wiring	D	A	W	A				
Insulation Resistance		A		A		A		A
Oil Dielectric		A		A		A		A
Oil Acidity		5 yr		5 yr		5 yr		5 yr
Filter and Reclaim Oil		NS		NS		NS		NS
Fire Protection	M	A						

D – routine daily inspection
W – routine weekly inspection
M – routine monthly inspection
Q – quarterly inspection
A – annual inspection
NS – not scheduled

GENERATOR SIZING FORM

Customer ____________ Project ____________ Analyst ____________ Date ________

I. APPLICATION DATA

Prime/Standby Power Gas/Diesel Fuel ________ Volts ________ Phase ________ Hz

II. LOADS **III. ENGINE SIZING**

A. Lighting Loads .... ________ kW

B. Other Non-Motor Loads .... ________ kW

C. Motors kW (Engine) = hp (Motor) x 0.746

		Nameplate Data				
Starting Sequence	hp	Nema Code	Reduced Voltage Starting Type	Acceptable Voltage Dip Percent	Motor Eff. (Chart 5)	Motor Efficiency (Chart 5)
1						kW
2						kW
3						kW
4						kW
5						kW

Total Motor Load ________ kW

Total Engine Load (A + B + C) ________ kW

IV. ENGINE SELECTION Model: ____________ Frame: ____________

Rating (With Fan): ________ kW ________ Hz ________ rpm

V. GENERATOR SIZING Start Sequence	Motor(s) 1	Motor(s) 2	Motor(s) 3
A. Starting kV•A (SKVA)			
1. Motor Ratings	______hp	______hp	______hp
2. NEMA Code	______	______	______
3. SKVA/hp (Use 6.0 if Code Letter Unknown)	______	______	______
4. SKVA/hp x Motor hp (A.1 x A.3)	______SKVA	______SKVA	______SKVA
B. Effective SKVA			
1. All Motors Running	0 kW	______kW	______kW
2. All Motors Running & Motor Being Started	______kW	______kW	______kW
3. $\frac{B.1}{B.2} \times 100$	0 %	______%	______%
4. Compensation for Motors Already Started (Chart 2)	1.0	______	______
5. Step A.4 x Step B.4	______SKVA	______SKVA	______SKVA
6. Reduced Voltage Factor (Chart 3) (use 1.0 if no starting aid used)	______	______	______
7. Effective SKVA = Step B.5 x B.6	______SKVA	______SKVA	______SKVA
8. Acceptable Voltage Dip (10, 20, 30%)	______%	______%	______%
C. Generator Selection (Chart 1)			
1. Frame	______	______	______
2. Rating	______kW	______kW	______kW
3. SKVA at Selected Voltage Dip	______	______	______

VI. GENERATOR SET SIZING

Select Largest Generator Set Model of Step IV and Step V.C.1.

Model: ______________ Frame: ________ Rating: ______kW Prime/Standby ______Hz ______rpm

CALCULATING AMPERES, HORSEPOWER, KILOWATTS & KILOVOLT AMPERES

TO FIND	DIRECT CURRENT	ALTERNATING CURRENT		
		SINGLE PHASE	TWO PHASE-FOUR WIRE	THREE PHASE
Amperes When HP Is Known	$\frac{HP \times 746}{E \times \%EFF}$	$\frac{HP \times 746}{E \times \%EFF \times PF}$	$\frac{HP \times 746}{E \times \%EFF \times PF \times 2}$	$\frac{HP \times 746}{E \times \%EFF \times PF \times 1.73}$
Amperes When KW Is Known	$\frac{KW \times 1000}{E}$	$\frac{KW \times 1000}{E \times PF}$	$\frac{KW \times 1000}{E \times PF \times 2}$	$\frac{KW \times 1000}{E \times PF \times 1.73}$
Amperes When KVA Is Known		$\frac{KVA \times 1000}{E}$	$\frac{KVA \times 1000}{E \times 2}$	$\frac{KVA \times 1000}{E \times 1.73}$
Kilowatts	$\frac{E \times I}{1000}$	$\frac{E \times I \times PF}{1000}$	$\frac{E \times I \times PF \times 2}{1000}$	$\frac{E \times I \times PF \times 1.73}{1000}$
Kilovolt-Amperes KVA		$\frac{E \times I}{1000}$	$\frac{E \times I \times 2}{1000}$	$\frac{E \times I \times 1.73}{1000}$
Horsepower	$\frac{E \times I \times \%EFF}{746}$	$\frac{E \times I \times \%EFF \times PF}{746}$	$\frac{E \times I \times \%EFF \times PF \times 2}{746}$	$\frac{E \times I \times \%EFF \times PF \times 1.73}{746}$

CHAPTER 8
METERS AND TESTING

METER TROUBLESHOOTING BASICS

Troubleshooting is the systematic elimination of the various parts of a system to locate a malfunctioning part. Proper tools and test equipment are essential to help troubleshoot problems quickly. The basic rules followed when using test instruments include:

- Always read and save the manufacturer's instructions for future reference.
- Always start with the highest scale available on a test instrument to prevent overloading due to unknown values.
- Always remove the component to be tested or disconnect the line voltage from the circuit before making any resistance measurements.
- Never try to use a test instrument beyond its rated capacity.
- Always close clamp-on instrument jaws tightly. All clamp-on instruments are designed to be clamped around one conductor.
- All leads must be insulated.
- All connections must be tight.
- Always check to ensure that any instrument fuses or batteries are in working condition.
- The needle on a clamp-on instrument should read in the upper half of the scale for greatest accuracy.
- Always apply basic rules of electrical theory when testing any circuit.

BASIC METER TYPES AND USES

Device	Measures	Unit of Measure	Typical Uses
Ammeter	Amount of electron flow in a circuit	Amperes (A)	Indicates amount of current a load or circuit is using
Conductivity Meter	Ability to conduct electricity	Mhos (usually Mhos/cm) or microsiemens (usually μ S/cm)	Measures the ability of a solution to conduct electricity
Frequency Meter	Number of electrical cycles per second	Hertz (Hz)	Indicates AC power line frequency in such applications as motor drives
Ohmmeter	Resistance to flow of electricity	Ohms (Ω)	Indicates a load, circuit, or component resistance before power is applied
Voltmeter	Amount of electrical pressure in a circuit	Volts (V)	Gives a voltage reading in applications such as battery chargers and power distribution systems
Wattmeter	Amount of electrical power in a circuit	Watts (W)	Gives power readings for distribution systems, heating elements, etc.

METER FUNCTIONS AND OPTIONS

The following represents what a meter should have to be both versatile and durable.

FUNCTIONS	OPTIONS
Analog Bar Graph Automatic Touch Hold Diode Test Frequency Measurement Min/Max Function Relative Mode	Audible Warning If Test Leads Are Connected Into Incorrect Jacks Fused Current Jack Water- or Chemical-Resistant Sealed Case

METER SPECIFICATIONS

Good	Specification	Better
2.9%	AC Voltage Accuracy	.7%
1.5%	DC Voltage Accuracy	.01%
600 V	Maximum AC/DC Voltage	1000 V
±0.9%	Resistance Accuracy	±0.2%
400 Hz	Frequency Range	20 kHz
Average	Multimeter Type	rms

METER TERMS

AC Coupling – Signal that passes an AC signal and blocks a DC signal. Used to measure AC signals that are riding on a DC signal.

AC/DC – Indicates ability to read or operate on alternating and direct current.

Accuracy Rating – Largest allowable error (in percent of full scale) made under normal operating conditions. The reading of a meter set on the 600V range with an accuracy rating of ±2% could vary ±12 Volts.

Ambient Temperature – Temperature of air surrounding a meter or equipment to test.

Ammeter Shunt – Low-resistance conductor that is connected in parallel with the terminals of an ammeter to extend the range of current values measured by the ammeter.

Amplitude – Highest value reached by a quantity under test.

Attenuation – Decrease in amplitude of a signal.

Autoranging – Function that automatically selects a meter's range based on signals received.

Average Value – Value equal to .637 times the amplitude of a measured value.

Battery Save – Enables a meter to shut down when battery level is too low or no key is pressed within a set time.

BNC – Coaxial-type input connector used by various meters.

Capture – Records and displays measured values.

Counts – Unit of measure of meter resolution. A 1999 count meter can display a measurement of $\frac{1}{10}$ of a volt when measuring 199V or less.

Decibel (dB) – Measurement that indicates voltage or power comparison in a logarithmic scale.

METER TERMS *(CONT.)*

Digits – Indication of the resolution of a meter. A 4½ digit meter can display four full digits and one half digit. The full digits display a number from 0 to 9. The half digit displays a 1 or is left blank.

Diode – Semiconductor that allows current to flow in only one direction.

Discharge – Removal of an electric charge.

Dual Trace – Feature that allows two separate waveforms to be displayed simultaneously.

Earth Ground – Reference point that is directly connected to ground.

Effective Value – Value equal to .707 of the amplitude of a measured quantity.

Freeze – Function that holds a waveform (or measurement) for closer examination.

Frequency – Number of complete cycles occurring per unit of time.

Glitch – Momentary spike in a waveform.

Glitch Detect – Function that increases the meter sampling rate to maximize the detection of the glitch.

Ground – Common connection to a point in a circuit whose potential is taken as zero.

Hold – Allows a meter to capture and hold a stable measurement.

Measuring Range – Minimum and maximum quantity that a meter can safely and accurately measure.

Noise – Unwanted extraneous electrical signals.

Overflow Overload – Condition of a meter that occurs when a quantity to be measured is greater than the quantity the meter can display or safely handle.

Peak – Highest value reached when measuring.

Peak-To-Peak – Highest and lowest voltage value of a waveform.

METER TERMS *(CONT.)*

Polarity – Orientation of the positive (+) and negative (-) side of direct current or voltage.

Pulse – Waveform that increases from a constant value, then decreases to its original value.

Pulse Train – Repetitive series of pulses.

Range – Quantities between two points or levels.

Recall – Stored information to be displayed.

Resolution – Sensitivity of a meter. A meter may have a resolution of 1 V or 1 mV.

Rising Slope – Part of a waveform displaying a rise in voltage.

Root-Mean-Square – Value equal to .707 of the amplitude of a measured value.

Sampling Rate – Number of readings taken from a signal every second.

Terminal Voltage – Voltage level that meter terminals can safely handle.

Trace – Displayed waveform that shows the voltage variations of the input signal as a function of time.

Trigger – Device which determines the beginning point of a wavelength.

Waveform – Pattern defined by an electrical signal.

Zoom – Allows a waveform (or part of waveform) to be magnified.

METER ABBREVIATIONS

AC = Alternating current or voltage

DC = Direct current or voltage

V = Volts

V_{avg} = Average voltage

V_{max} = Peak voltage

V_{p-p} = Peak-to-peak voltage

V_{rms} = Root-mean-square voltage

mV = Millivolts

kV = Kilovolts

A = Amperes

mA = Milliamperes

μA = Microamperes

W = Watts

Hi-Z = High input impedance

dB = Decibel

dBV = Decibel volts

dBW = Decibel watts

kΩ = Kilohms

MΩ = Megohms

Hz = Hertz

kHz = Kilohertz

μF = Microfarads

nF = Nanofarads

°F = Degrees Fahrenheit

°C = Degrees Celsius

RPM = Revolutions per minute

COM = Common

OL = Overload

T = Time

LSD = Least significant digit

MAX = Maximum

MIN = Minimum

AVG = Average

TRIG = Trigger

METER SYMBOLS

~	AC
	DC
	AC or DC
+	Positive
−	Negative
	Ground
±	Plus or Minus
	Diode
	Diode test
<	Less than
>	Greater than
△	Increase setting
▽	Decrease setting
	See service manual
	Double insulation
	Fuse
	Battery
	Hold
	Audio beeper
	Capacitor
%	Percent
▷	Move right
◁	Move left
	No (do not use)

METER SYMBOLS *(CONT.)*

○	Switch position OFF (power)
|	Switch position ON (power)
	Manual range mode
	Warning: Dangerous or high voltage that could result in personal injury
	Caution: Hazard that could result in equipment damage or personal injury
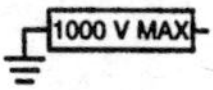	Terminals must not be connected to a circuit with higher-than-listed voltage
Δ	Relative mode - displayed value is difference between present measurement and previous stored measurement
Ω	Ohms resistance
	Meter display light

HIGH VOLTAGE FIELD ACCEPTANCE TEST

Rated Voltage Phase to Phase	dc Hi-Pot Test (15 Minutes)		dc Hi-Pot Test	
	Wall-mils	kV	Wall-mils	kV
5000	90	25	115	35
8000	115	35	140	45
15000	175	55	220	65
25000	260	80	320	95
28000	280	85	345	100
35000	345	100	420	125
46000	445	130	580	170
69000	650	195	650	195

Note: If the leakage current quickly stabilizes, the duration may be reduced to 10 minutes.

CONNECTIONS FOR A BELTED CORE-LOSS AND FRICTION TEST

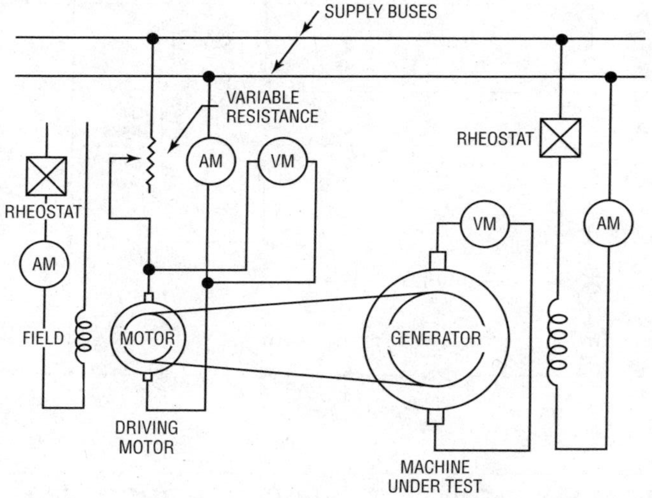

CONNECTIONS FOR KAPP'S LOADING-BACK TEST

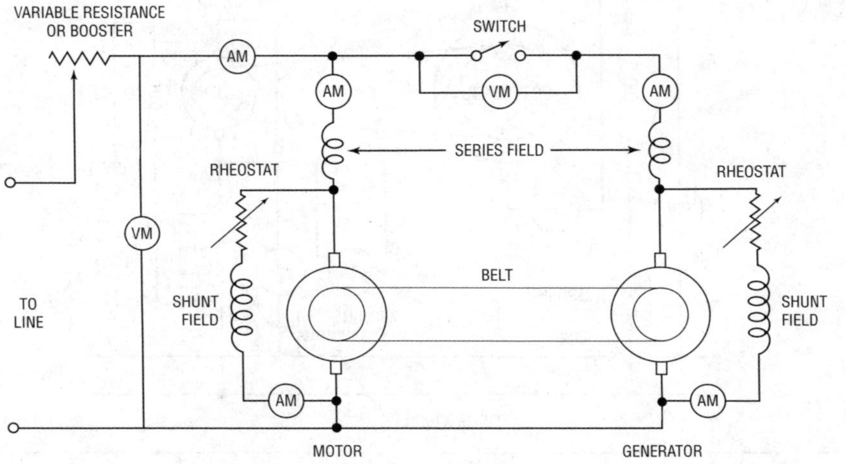

CONNECTIONS FOR HOPKINSON'S LOADING-BACK TEST

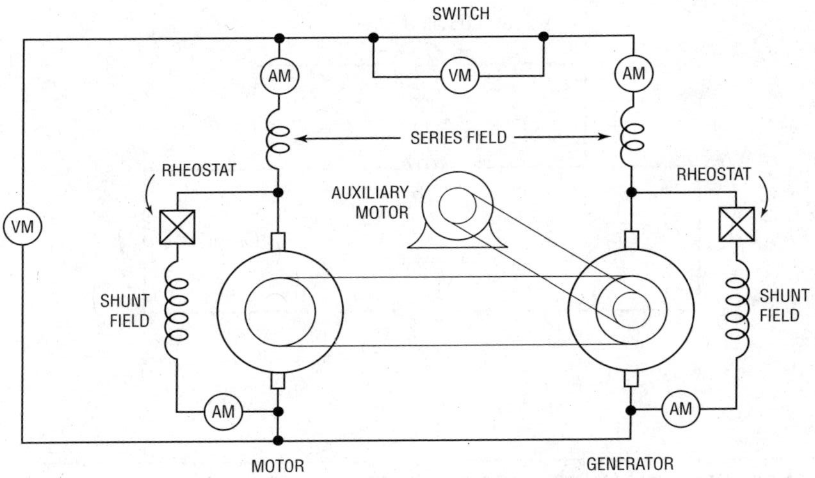

AN AMMETER AND VOLTMETER CONNECTED INTO A CIRCUIT

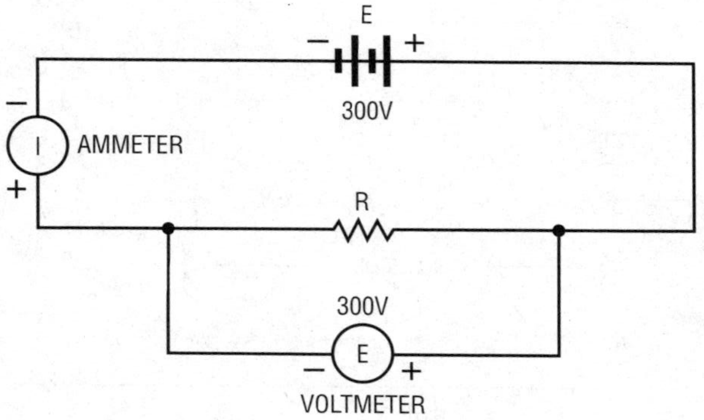

METER AND HOW IT IS USED WITH A 2 OR 3 WIRE SUPPLY

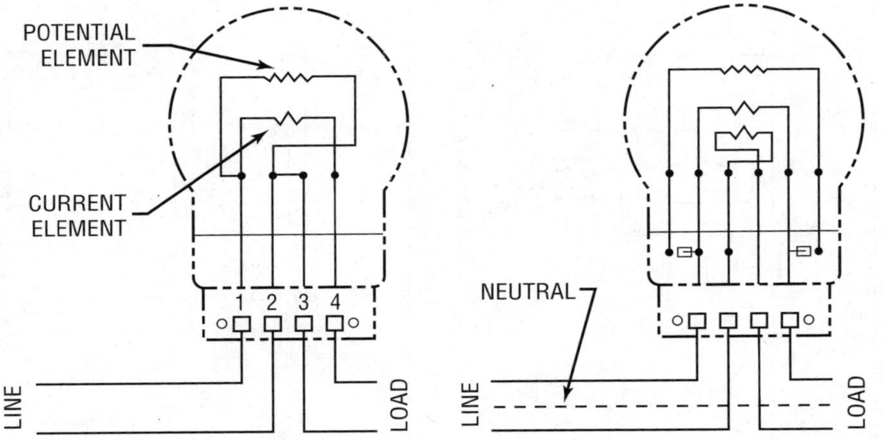

MEASURING POWER

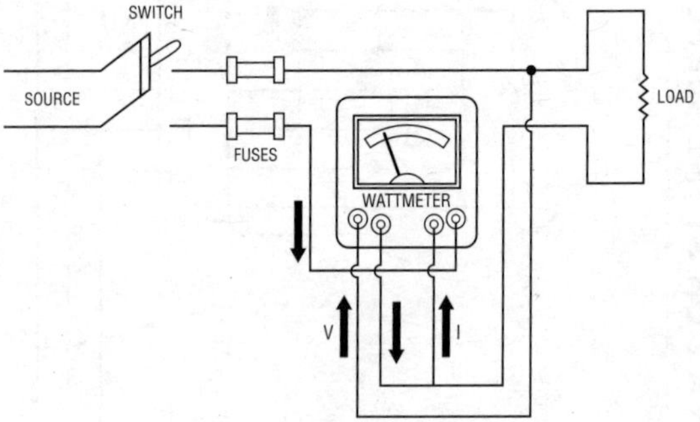

With wattmeter

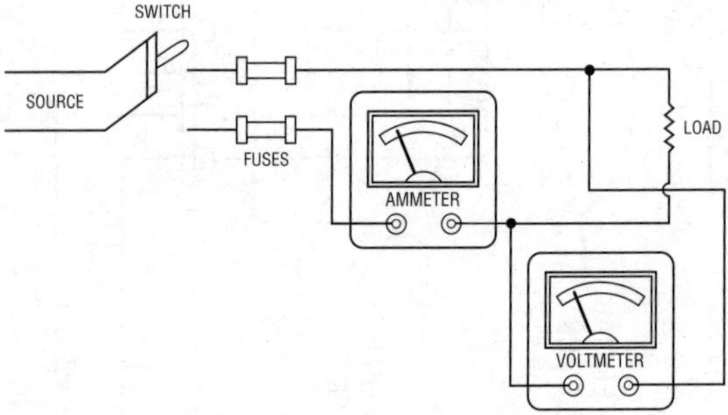

With ammeter and voltmeter

MEASURING POWER (cont.)

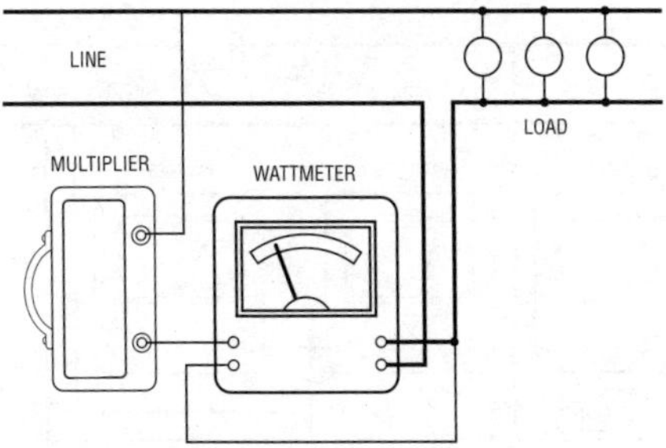

Wattmeter connection at a single-phase circuit using a multiplier in the meter potential circuit.

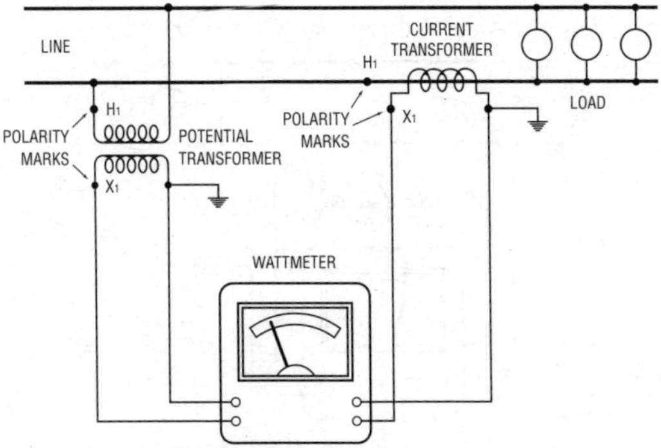

Wattmeter connection in a single-phase circuit using current and potential transformers.

MEASURING POWER (cont.)

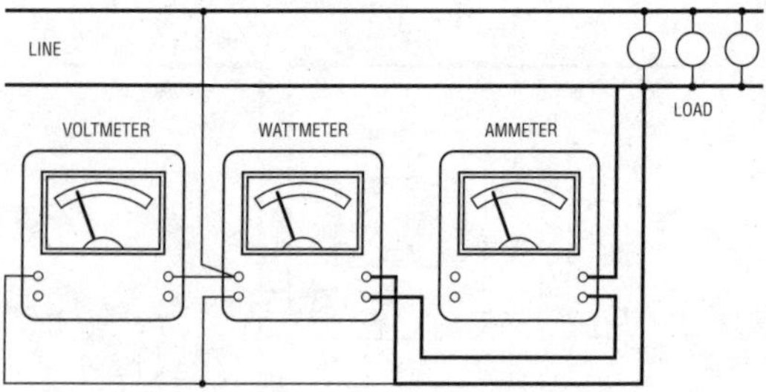

Connection of a wattmeter, voltmeter, and ammeter to a single-phase circuit.

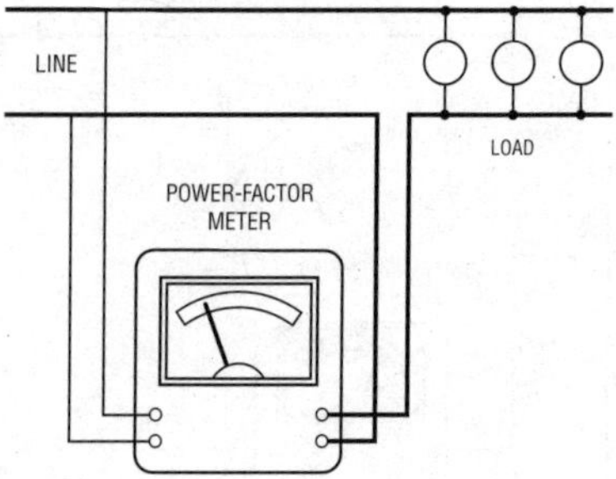

Connection of a power-factor meter to a single-phase circuit.

MEASURING POWER (cont.)

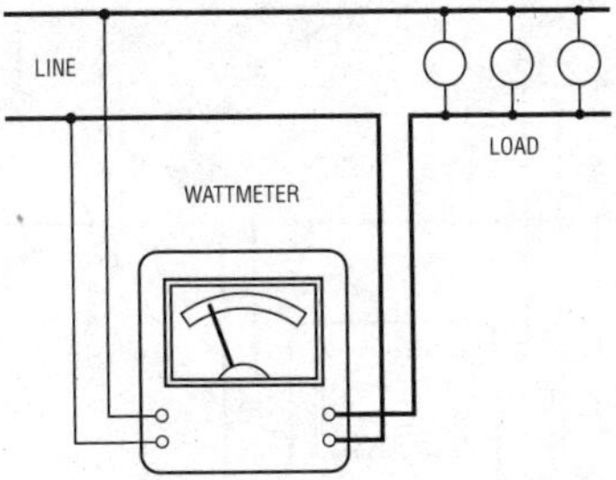

Wattmeter connection to a single-phase circuit where the meter measures the load power plus the loss in its own current coil.

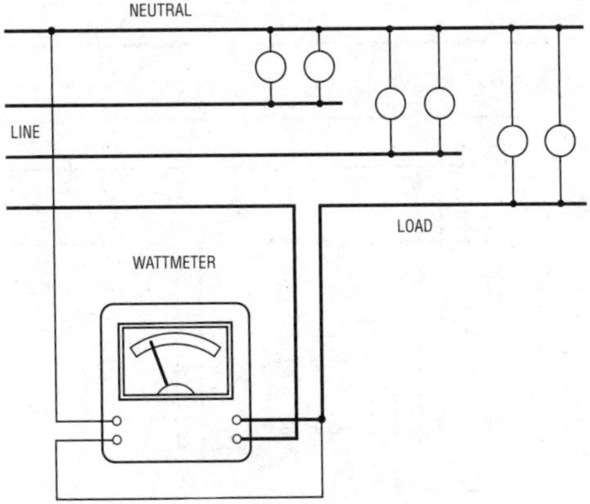

Connection of a single-phase wattmeter to a balanced four-wire, three-phase circuit.

MEASURING POWER (cont.)

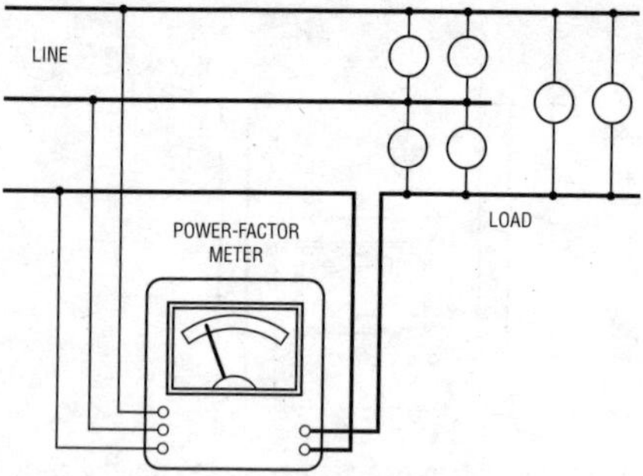

Connection of a power-factor meter to a three-wire, three-phase circuit.

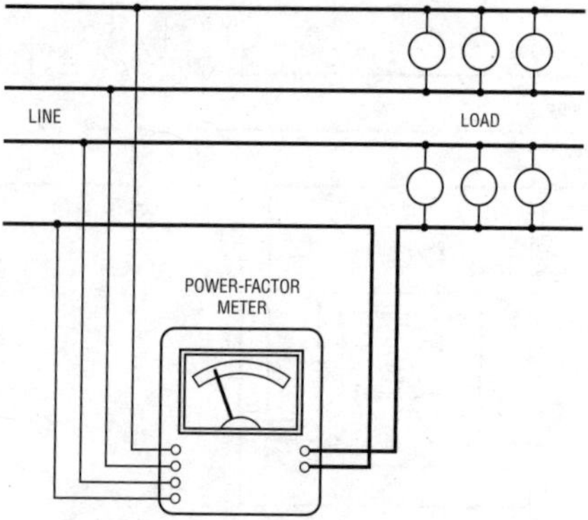

Connection of a power-factor meter to a four-wire, three-phase circuit.

MEASURING POWER (cont.)

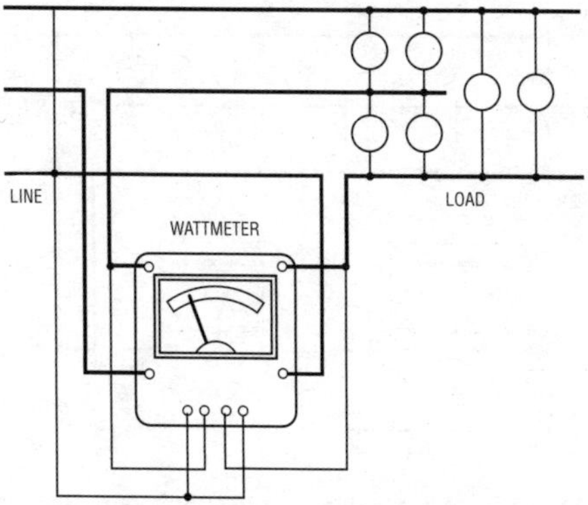

Connection of a polyphase wattmeter to a three-wire, three-phase circuit.

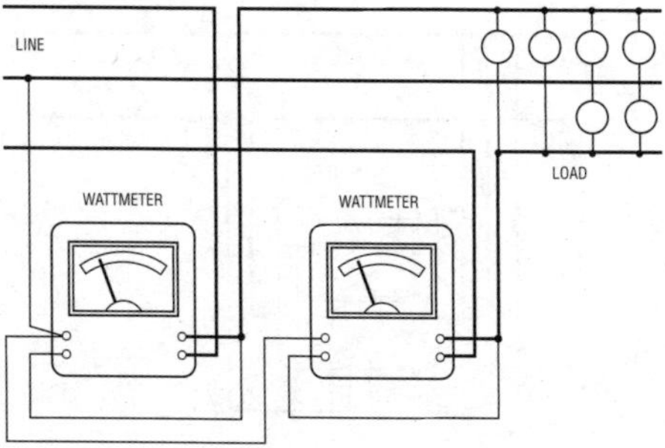

Connection of two wattmeters to a three-phase circuit having balanced or unbalanced voltages or load.

MEASURING POWER (cont.)

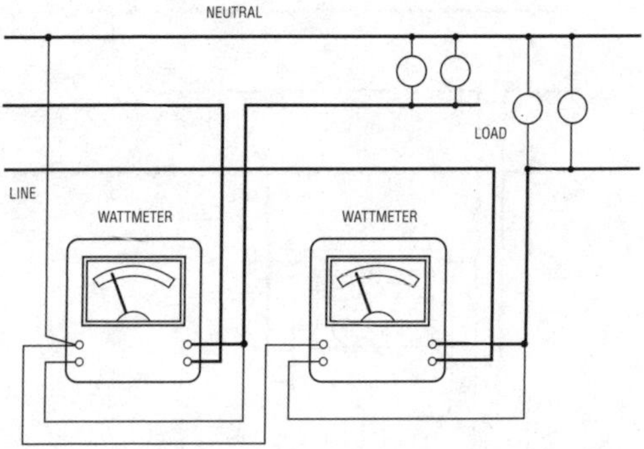

Connection of two wattmeters to a three-wire, two-phase circuit having a balanced or unbalanced load.

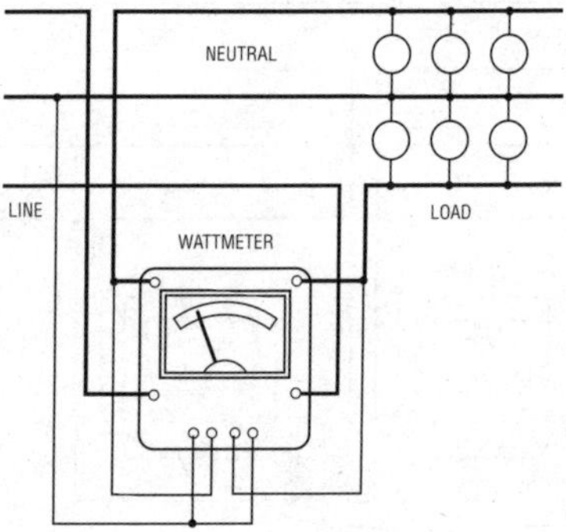

Connection of a polyphase wattmeter to a three-wire, two-phase circuit having balanced or unbalanced voltages or load.

MEASURING CURRENT

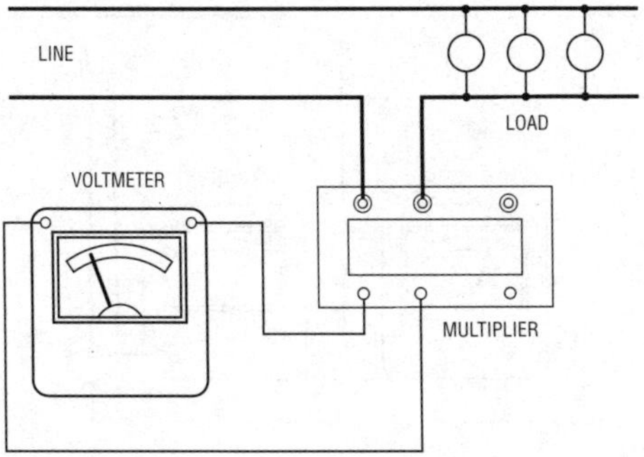

DC ammeter connected to a shunt for measuring DC current.

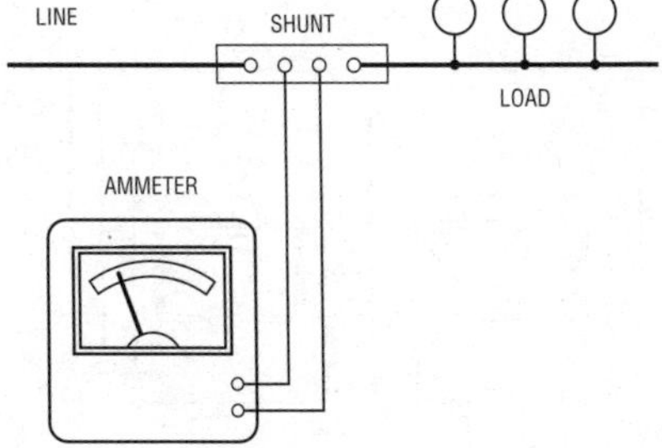

DC ammeter connected to a shunt to extend the range of the meter.

MEASURING VOLTAGE

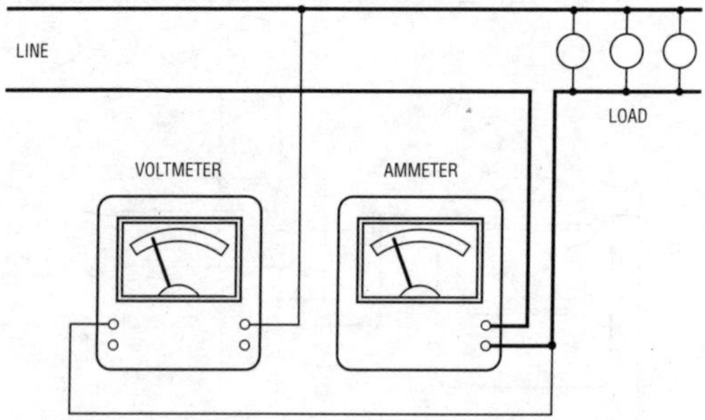

Voltmeter and ammeter connection in which the voltmeter reads the true voltage across the load.

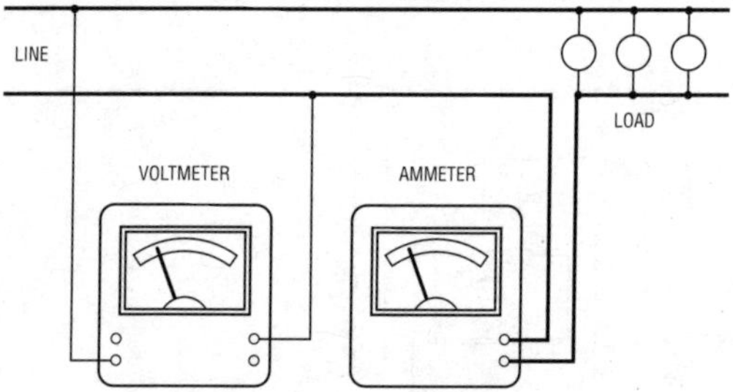

Voltmeter and ammeter connection in which the voltmeter reads the true line voltage.

MEASURING VOLTAGE (cont.)

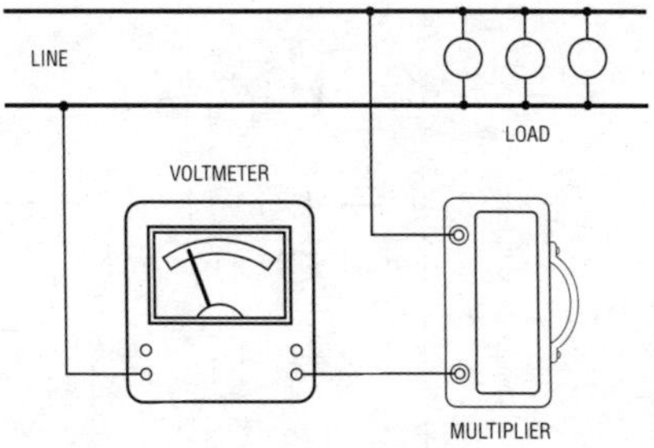

AC voltmeter connection using an external multiplier to extend the meter range.

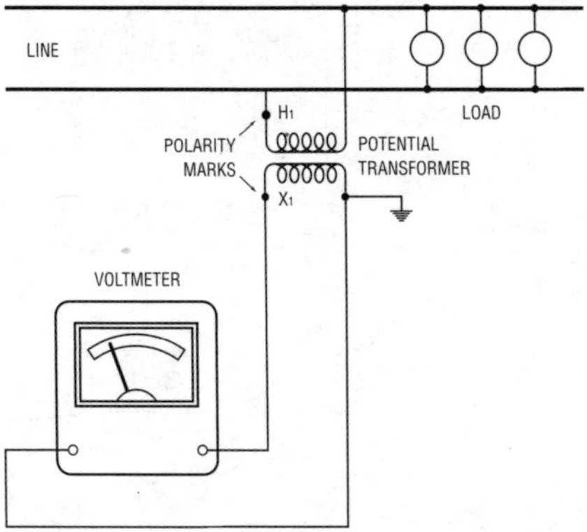

AC voltmeter used with a potential transformer.

VOLTAGE MEASUREMENTS OF A 2-LAMP, RAPID-START, SERIES-TYPE, ELECTROMAGNETIC FLUORESCENT BALLAST

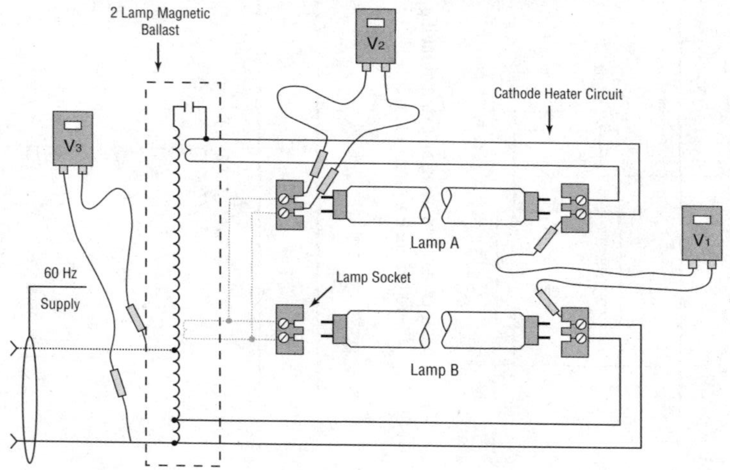

CHAPTER 9
MOTOR MAINTENANCE

Preventive maintenance is performed to keep electrical motor and power transmission equipment running with little or no downtime. In the past, the job of the maintenance department was almost always to repair broken equipment and install new equipment.

Because of the high cost of labor, installing new motors is still a top priority of maintenance departments. However, inexpensive monitors are available that can monitor an electrical system for voltage or phase unbalances, voltage losses, phase reversals, over or undervoltages, currents, temperatures, loss of a pump's prime and other conditions that may be signs of major problems.

A preventive maintenance program includes inspection, cleaning, tightening, adjusting and lubricating, keeping equipment dry, and electronically monitoring power circuits. The purpose of a preventive maintenance program is to:

- **Maintain equipment to ensure uninterrupted operations at the highest efficiency for as long as possible.**
- **Protect equipment from dirt, dust, moisture, corrosion, and electrical and mechanical overloads.**
- **Maintain records of all maintenance work to establish future needs and priorities.**

LOCATING CIRCUITS

Repairs and upgrades to circuits frequently require the circuit to be identified in the panel that feeds it, even if only to disconnect power. This can be a problem, especially in a poorly-marked, crowded panel. Turning circuits off one-by-one in a panel is not a good option, as some of the circuits may feed sensitive devices.

A clamp-on ammeter and a flashing lamp can be used to solve this problem quickly and easily. Plug the flashing lamp into any receptacle outlet on the circuit you need to identify, then check the circuits in the panel. The circuit whose current rises and falls along with the flashing of the lamp is the one.

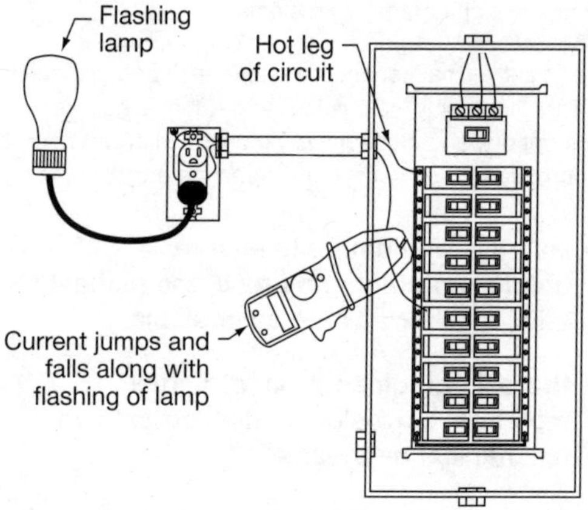

CHECKING CAPACITORS

Capacitors burn out after a limited period of use, and should be checked regularly in applications where burnout would have serious consequences.

The first method in checking a capacitor is to inspect it visually for cracks, bulges, or leakage. this provides a fast check for obvious problems, but will not find internal flaws.

A more thorough check is made by using an ohmmeter. (A typical multimeter will have ohmmeter capabilities.) But before this can be done, the capacitor must be discharged. This is done by removing it from the circuit, then connecting a resistor across its leads. A 20 kΩ, 5 watt resistor will usually work, and will discharge the capacitor in 5 – 10 seconds.

Once the capacitor is discharged, the ohmmeter can be connected across the capacitor's terminals. If the capacitor is good, the meter will initially show zero resistance, then slowly move to infinity. If the capacitor is shorted, the meter will show zero, and will not change. If the capacitor is open, the meter will show infinity, and will not change.

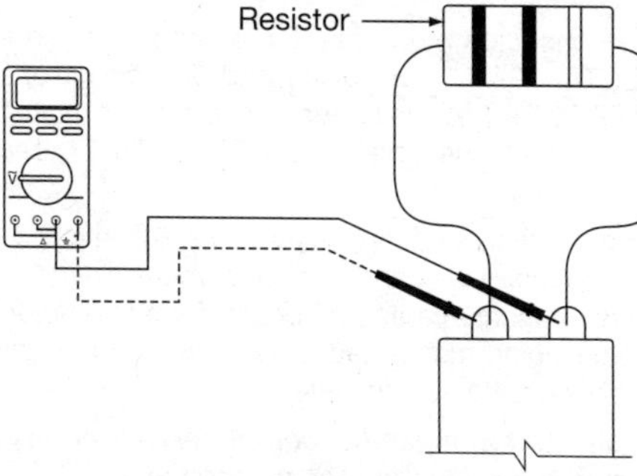

UNMARKED 3Φ INDUCTION MOTORS

When a motor is in operation for a long period of time, it may cause the markings of the external leads to become defaced. This can also occur to a rebuilt motor that has been in the maintenance shop for awhile. To ensure proper operation, each motor lead must be remarked.

The most common 3ϕ motors are the single-voltage, three-lead motor and the dual-voltage, nine-lead motor. Both may be internally connected in a wye or delta configuration.

The three leads of a single-voltage, 3ϕ, three-lead motor can be marked as T1, T2, and T3 in any order. The motor can be connected to the rated voltage and allowed to run. T1 and T3 may be interchanged if the rotation is in the wrong direction.

WYE OR DELTA CONNECTION

A multimeter can be used to determine whether a dual-voltage motor is internally connected in a wye or delta configuration.

A wye-connected motor has three circuits of two leads each (T1-T4, T2-T5, and T3-T6) and one circuit of three leads (T7-T8-T9). A delta-connected motor has three circuits of three leads each (T1-T4-T9, T2-T5-T7, and T3-T6-T8).

Determine the winding circuits (T1-T4, etc.) on an unmarked motor by connecting one meter lead to any motor lead and temporarily connecting the other meter lead to each remaining motor lead. A resistance reading other than infinity indicates a complete circuit.

Caution: Motor must be completely disconnected from circuit before testing for resistance.

WYE-CONNECTED MOTOR

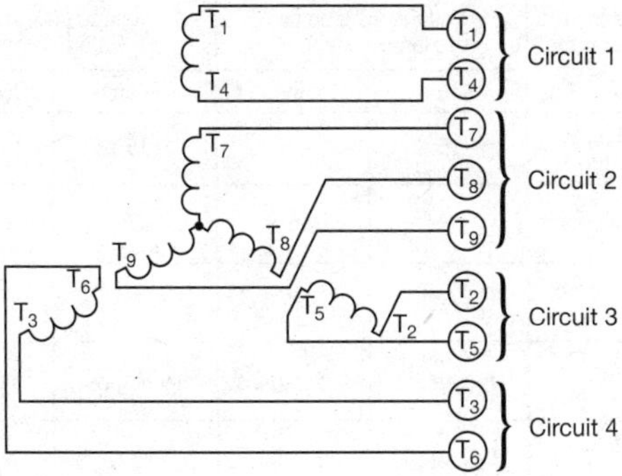

DELTA-CONNECTED MOTOR

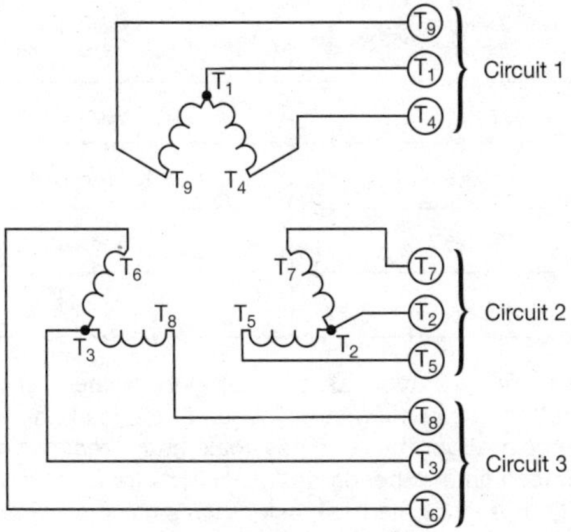

DC MOTOR PERFORMANCE CHARACTERISTICS

Performance Characteristics	Voltage 10% below Rated Voltage		Voltage 10% above Rated Voltage	
	Shunt	Compound	Shunt	Compound
Starting Torque	–15%	–15%	+15%	+15%
Speed	–5%	–6%	+5%	+6%
Current	+12%	+12%	–8%	–8%
Field temperature	Increases	Decreases	Increases	Increases
Armature temperature	Increases	Increases	Decreases	Decreases
Commutator temperature	Increases	Increases	Decreases	Decreases

MAXIMUM ACCELERATION TIME

Frame Number	Maximum Acceleration Time
48 and 56	8 Seconds
143-286	10 Seconds
324-326	12 Seconds
364-505	15 Seconds

The longer a motor takes to accelerate, the higher the temperature rise in the motor. The larger the load, the longer the acceleration time. The maximum recommended acceleration time depends on the motor's frame size. Large motor frames dissipate heat faster than small motor frames.

AC VOLTAGE VARIATION CHARACTERISTICS

Performance Characteristics	10% Above Rated Voltage	10% Below Rated Voltage
Starting current	+10% to +12%	–10% to –12%
Full-load current	–7%	+11%
Motor torque	+20% to +25%	–20% to –25%
Motor efficiency	Little change	Little change
Speed	+1%	–1.5%
Temperature rise	–3°C to –4°C	+6°C to +7°C

Motor performance is affected when the supply voltage varies from a motor's rated voltage. A motor operates satisfactorily with a voltage variation of ±10% from the voltage rating listed on the motor nameplate.

AC FREQUENCY VARIATION CHARACTERISTICS

Performance Characteristics	5% Above Rated Frequency	5% Below Rated Frequency
Starting current	–5% to –6%	+5% to +6%
Full-load current	–1%	+1%
Motor torque	–10%	+11%
Motor efficiency	Slight increase	Slight decrease
Speed	+5%	–5%
Temperature rise	Slight decrease	Slight increase

Motor performance is affected when the frequency varies from a motor's rated frequency. A motor operates satisfactorily with a frequency variation of ±5% from the frequency rating listed on the motor nameplate.

PHASE UNBALANCE AND TEMPERATURE RISE

The loads of three-phase power systems are balanced during installation. An unbalance can begin if additional single-phase loads are added to the system. As these additional loads are energized, one or two of the three-phase lines will begin to carry more or less of the load of the system. This unbalance causes the three-phase lines to move out-of-phase so the lines are no longer 120 electrical degrees apart.

Phase unbalance causes three-phase motors to run at temperatures higher than their listed ratings. The graph below illustrates the relationship between phase inbalance and motor temperature in percentages.

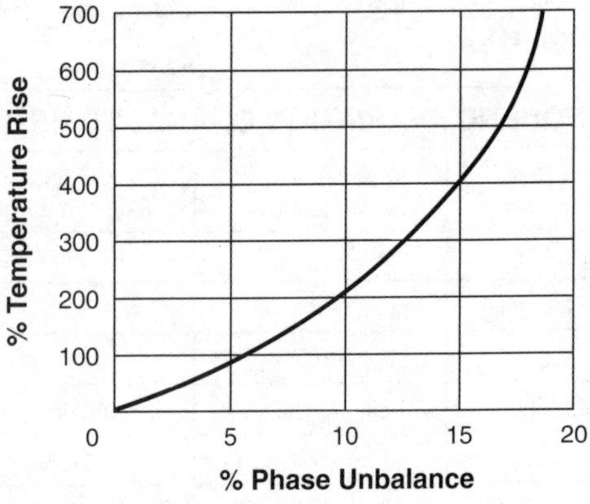

The greater the phase unbalance, the greater the temperature rise. High temperatures produce insulation breakdown and other related problems such as motor inefficiency.

PHASE UNBALANCE DERATING FACTOR

A three-phase motor operating in an unbalanced circuit cannot deliver the horsepower it is rated for. For example, a phase unbalance of 5% will cause a motor to work at 75% of its rated power. This results in the motor having to be derated. The graph below illustrates this relationship.

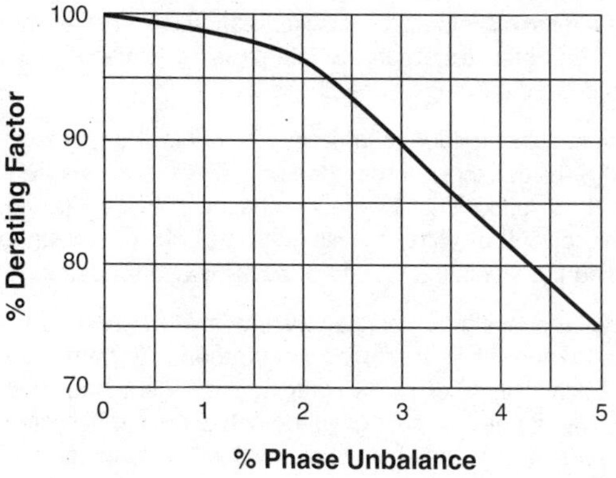

A motor operating on a circuit which has phase unbalance must be derated and will not perform its intended function of work as economically as was originally designed.

It is imperative that the electrical maintenance department check the loads of a three-phase system whenever additions to that system occur.

SINGLE-PHASING CONDITION

Single-phasing is when a three-phase motor is only operating on two phases because one phase has been lost. Single-phasing occurs when one of the lines leading to the motor does not deliver voltage. Single-phasing is the absolute maximum condition of voltage unbalance.

Single-phasing can be distinguished from voltage unbalance by the degree of damage. Voltage unbalance causes less discoloration (blackening), usually over more coils with little distortion. Single-phasing causes severe distortion and burning to one phase coil.

Severe blackening of one delta winding or two wye windings of the three three-phase windings occurs when a motor has failed due to single phasing. The coil(s) that experienced the voltage loss indicate obvious damage including the insulation on one or possibly two windings.

Possible causes of single-phasing are when one fuse blows, when the switching equipment undergoes a mechanical failure or when lightning blows out one of the lines. Single-phasing can go undetected on most systems because a motor running on two-phase will continue to run in most applications until it burns out.

In most cases, measuring a motor's voltage does not detect a single-phasing condition. The open winding in the motor generates a voltage almost equal to the phase voltage that is lost. When that happens, the open winding acts as the secondary of a transformer, and the two windings connected to power act as the primary. The conditions for single-phasing can be reduced by using correct heater sizes and by using proper size dual-element fuses.

IMPROPER PHASE SEQUENCE (PHASE REVERSAL)

Improper phase sequence is the changing of the sequence of any two phases in a three-phase motor control circuit. Improper phase sequence reverses the motor rotation causing severe mechanical damage and possible injuries to personnel.

Phase reversal can occur when modifications are made to a power distribution system or when maintenance is performed on electrical conductors or switching equipment. Again, care must be taken to properly identify and mark all components of any electrical system undergoing those types of procedures.

In addition, the NEC® requires phase reversal protection on many types of transportation equipment such as escalators, elevators, etc.

VOLTAGE SURGE

A *voltage surge* is a higher-than-normal voltage that temporarily exists on one or more of the power lines. Lightning is a major cause and the surge on a power line can come from a direct hit or induced voltage. Lightning energy can move in both directions on power lines and this traveling surge causes a large voltage rise quickly. The large voltage is impressed on the first couple of turns of the windings, destroying the insulation and causing the motor to burn out. The rest of the windings will appear normal, with very little or no damage. Properly sized (voltage rating) lightning arresters combined with a connection to a good ground will result in maximum voltage surge protection.

For further protection, surge protectors should be installed on the equipment of the system.

VOLTAGE PROBLEMS

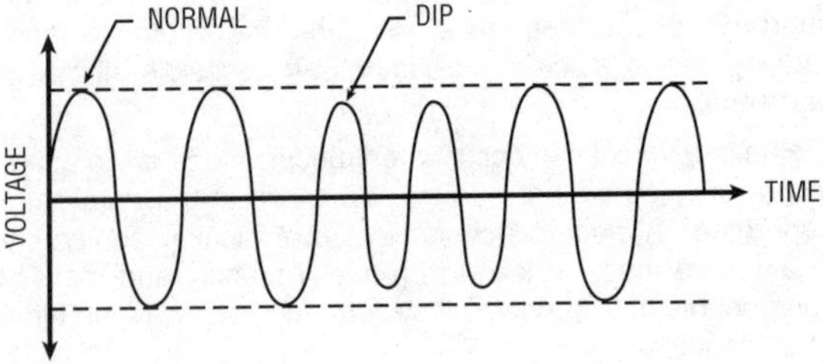

Voltage Dip

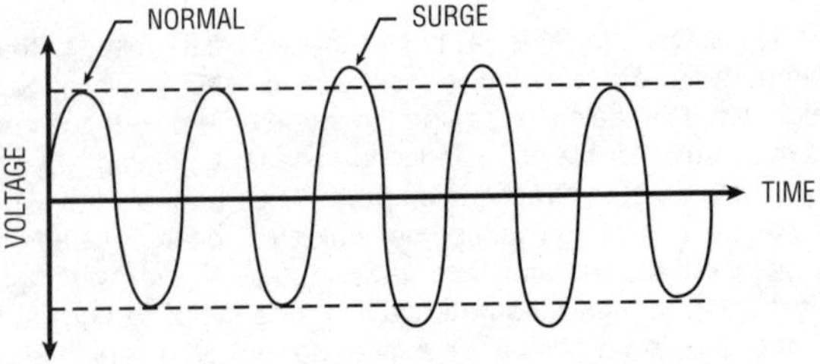

Voltage Surge

VOLTAGE VARIANCE

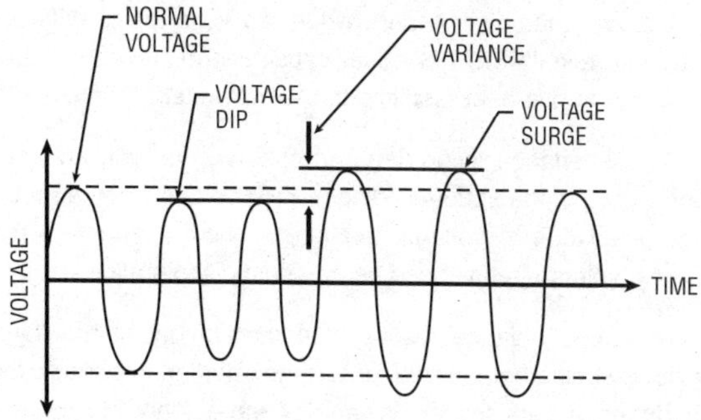

ACCEPTABLE AC LOAD VOLTAGE RANGES (60Hz)

Rating	Minimum	Maximum
24	22	26
115	109	121
120	114	126
208	197	218
230	218	242
240	228	252
277	263	291
460	436	484
480	456	504

VOLTAGE UNBALANCE

Voltage unbalance occurs when the voltages at different motor terminals are not equal causing one or more motor windings to overheat resulting in thermal deterioration.

A maintenance technician can observe the blackening of one or two of the stator windings which occurs when a motor has failed due to voltage unbalance. The winding with the largest voltage unbalance has the most discoloration.

Voltage unbalance results in a current unbalance. Line voltage should be checked for voltage unbalance periodically or during a service call. Whenever more than 2% voltage unbalance is measured, the following actions should be utilized:

- **The surrounding power system should be examined for excessive loads connected to any one line.**
- **Adjust the excessive load on the motor if the voltage unbalance cannot be corrected.**
- **Change the motor rating by installing an oversized motor if the voltage unbalance cannot be corrected.**
- **Notify your local electric utility of the problem if all else fails.**

FINDING VOLTAGE UNBALANCE

1. Measure the voltage between each incoming power line. Take the readings as follows:

   L1 to L2, L1 to L3 and L2 to L3

2. Add the voltages, then divide by 3 to find the voltage average.

3. Find the voltage deviation by subtracting the voltage average from the voltage with the largest deviation.

4. Apply the following formula to find the voltage unbalance percentage:

$$V_u = \frac{V_d}{V_a} \times 100$$

V_u = voltage unbalance (%)

V_d = voltage deviation (in V)

V_a = voltage average (in V)

100 = percentage constant

CALCULATING VOLTAGE UNBALANCE

The following example illustrates the calculation of the voltage unbalance percentage in a feeder system with the following voltage readings:

L1 to L2 = 444V;

L1 to L3 = 458V and

L2 to L3 = 472V.

ADD VOLTAGES

444V
458V
472V
1374V

FIND VOLTAGE AVERAGE

$$V_a = \frac{V}{3}$$

$$V_a = \frac{1374}{3}$$

$$V_a = 458V$$

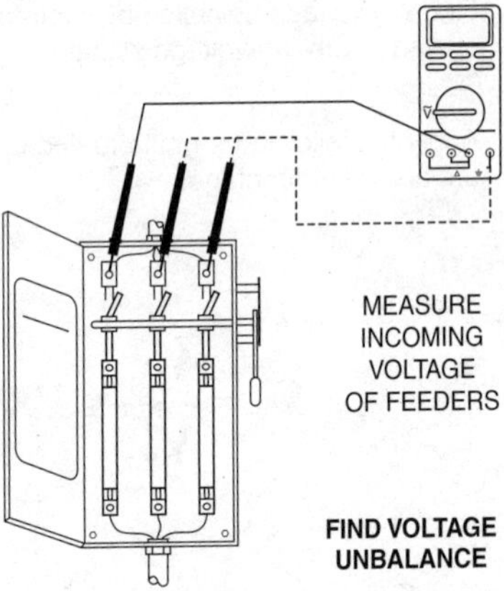

MEASURE INCOMING VOLTAGE OF FEEDERS

FIND VOLTAGE DEVIATION

$$V_d = V - V_a$$

$$V_d = 472 - 458$$

$$V_d = 14V$$

FIND VOLTAGE UNBALANCE

$$V_u = \frac{V_d}{V_a} \times 100$$

$$V_u = \frac{14}{458} \times 100$$

$$V_u = .0306 \times 100$$

$$V_u = 3.06\%$$

MOTOR OVERCYCLING

Overcycling is the process of a motor turning on and off repeatedly. Motor starting current is normally 6 to 8 times the full-load running current. The majority of motors are not designed to start more than 10 times per hour. Overcycling begins when a motor is at its operating temperature and still cycles on and off further increasing the temperature and damaging the insulation of the motor.

When your application requires a motor to be cycled often, the following tips may be useful:

- **Utilize a motor with a 50°C rise instead of 40°C.**
- **Utilize a motor with a 1.25 or 1.35 service factor in place of a motor with a factor of 1.00 or 1.15.**
- **Force air over the motor to dissipate the additional heat.**

IMPROPER VENTILATION

Every motor produces heat which must be removed to prevent damage to the insulation. Motors are designed with passages that allow free airflow through the motor removing the heat. Restricting the airflow causes the motor to operate at higher temperatures.

Airflow can be restricted by the accumulation of grease, sludge, grass trimmings, vermin, dirt, etc. Airflow is also restricted if a motor becomes coated with oil from leaks or excessive lubrication. If a motor is placed in an enclosed area it can overheat due to the recirculation of heated air. Adding vents will allow the heated air to dissipate, thus keeping the motor operating at a lower temperature.

EXCESSIVE HEAT

One of the biggest causes of motor problems and failures is excessive heat. Excessive heat will destroy the insulation which, in turn, shorts the windings and renders the motor non-operational.

Insulation life is shortened as the heat in a motor increases beyond the insulation's temperature rating. The temperature rating of motor insulation is listed by classes and is illustrated in the chart below. A motor nameplate almost always lists the insulation class.

MOTOR INSULATION CLASS

Class	°F	°C
A	221	105
B	266	130
F	311	155
H	356	180

Excessive heat is normally caused by the following conditions:

- **Incorrect motor sizing for the intended application.**
- **Improper ventilation from various sources.**
- **Excessive load on the motor.**
- **Excessive friction from vibration due to improper installation and/or alignment with other mechanical devices coupled to the motor shaft.**
- **Electrical problems such as overloads, phase unbalances, voltage surges or voltage unbalances.**

MOTOR OVERLOADS

The term *overload* refers to a condition that results from applying an excessive load to a motor. Motors try to drive the connected load when power is applied. Obviously, the larger the load, the more power is required to drive the load. Every motor that is manufactured has a rated limit to the load they are able to drive.

For example, a 10 HP, 460 V, 3 Φ motor should draw no more than 14 amps of current. If a motor begins to draw more than its rating, overloading will begin. Current readings must be taken to determine if an overload problem exists. The chart below illustrates meter readings and their relationship to the motor's listed rating from the nameplate.

Current Rating Of Motor	Meter Readings		
	Fully Loaded Motor	**Underloaded Motor**	**Overloaded Motor**
20 A	20 A	12 A	22 A
Nameplate Rating	95 to 105% of Rating	0 to 95% of Rating	105% + of Rating

The even blackening of all of the motor windings occurs when a motor has failed due to overloading. This even discoloration is due to the slow destruction of the windings over a long time period. Normally, no visible damage to the insulation of the motor can be detected.

If a motor is protected properly, overloads should not become a problem. Using correctly-sized heaters in the motor starter will assure that an overload is removed from the motor before any damage occurs.

MEGOHMMETER CONNECTIONS

A megohmmeter (frequently called a *megger*) is used to check insulation strength with a high voltage, usually between 50 and 5000 volts. A megohmmeter may be used to test between circuit conductors or windings, or may test between the conductor or winding to ground.

Since insulation integrity varies over time (and may also vary with other conditions), megger testing is recommended every six months for critical circuits or equipment.

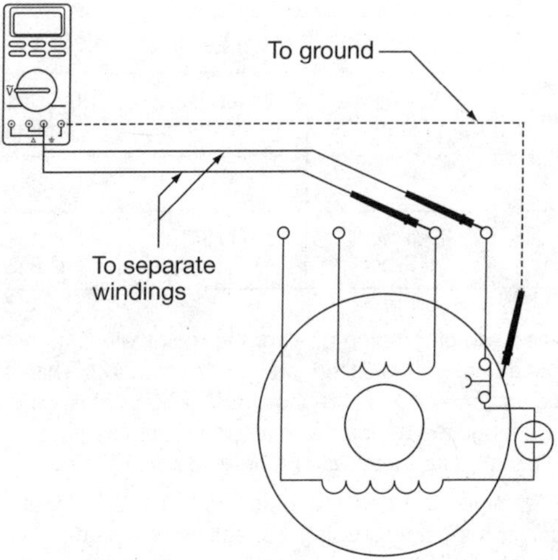

OHMMETER CONNECTIONS

An ohmmeter measures the resistance of windings and components in a motor circuit.

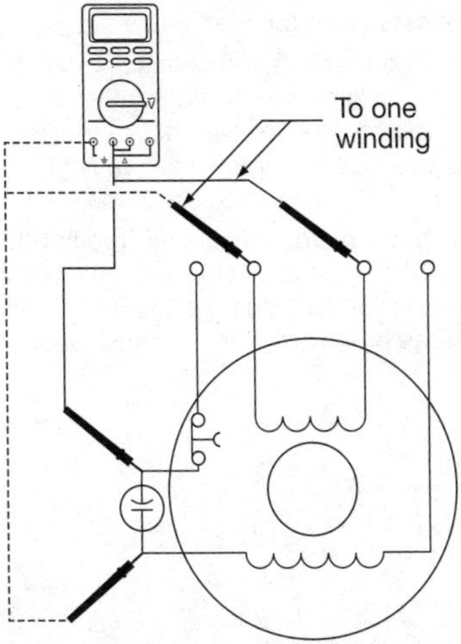

Typical Resistance Values at 40°C

Motor Voltage	Minimum Acceptable Resistance
Up to 208 V	100,000 Ω
208-240 V	200,000 Ω
240-600 V	300,000 Ω
600-1000 V	1 MΩ
1000-2400 V	2 MΩ
2400-5000 V	3 MΩ

INSULATION SPOT TESTING

Insulation spot testing verifies insulation quality over the life of the motor. Insulation spot testing should be done every six months using the method below:

A. Measure the resistance of each winding lead to ground. Record the readings after 60 seconds. Then record the lowest meter reading on an insulation spot test graph if all readings are above the minimum acceptable resistance.

B. Connect the motor windings to ground through a 5 kΩ, 5 W resistor. The winding should be connected for 10 times the motor testing time to discharge energy stored in the insulation.

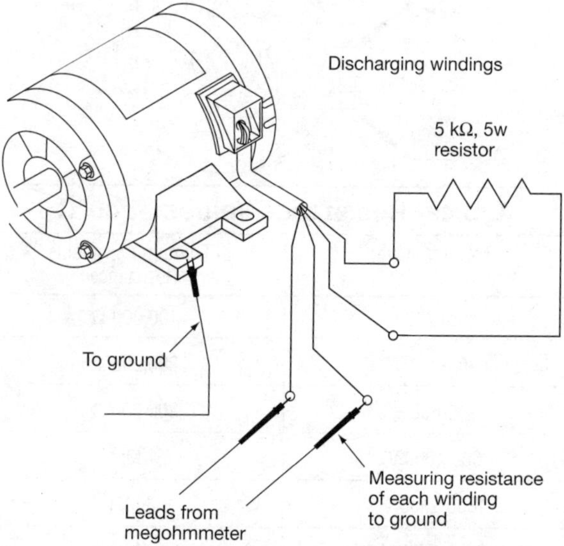

DIELECTRIC ABSORPTION TESTING

A *dielectric absorption test* checks the absorption characteristics of moist or cracked insulation and is performed over a 10-minute period.

A. Using a megohmmeter, measure the resistance of each winding lead to ground. Record the lowest meter reading on a dielectric absorption test graph if all readings are above the minimum acceptable resistance. Record the readings every 10 seconds for the first minute and every minute thereafter for 10 minutes.

B. Discharge the motor windings.

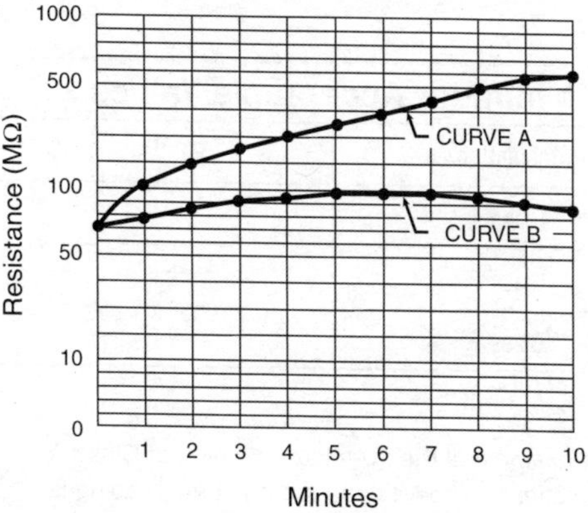

The good insulation is illustrated by Curve G above which shows a gradual increase in resistance levels over time. Cracked or moist insulation (as shown by Curve MC) illustrates relatively constant resistance levels over time.

Note: Service the motor if any reading does not meet the minimum acceptable resistance.

POLARIZATION INDEX VALUES

The polarization index indicates the overall condition of motor insulation. Excessive moisture and/or cracks in the insulation are indicated by a low polarization index value.

A polarization index is obtained by dividing the value of the 10-minute reading by the value of the 1-minute reading from a dielectric absorption test.

$$\text{P.I.} = \frac{\text{10-minute M}\Omega\text{ Reading}}{\text{1-minute M}\Omega\text{ Reading}}$$

Minimum Acceptable Values	
Insulation	P.I. Value
Class A	1.5
Class B	2.0
Class F	2.0

For example, if the 10-minute reading of Class A insulation is 120 MΩ and the 1-minute reading is 90 MΩ, the polarization index is 1.33 showing the insulation contains excessive moisture or cracks.

INSULATION STEP VOLTAGE TESTING

An *insulation step voltage test* creates electrical stress on internal insulation cracks to reveal damage not found during other insulation tests.

A. Use a megohmmeter setting of 500 Volts to measure the resistance of each winding lead to ground. Observe each reading after one minute and record the lowest reading.

B. Place the meter leads on the winding that has the lowest reading, set the megohmmeter at 1000 Volts, then take a reading starting at 1000 Volts and ending at 5000 Volts, at 500 Volt intervals. Record each reading after one minute has elapsed.

C. Discharge the motor windings.

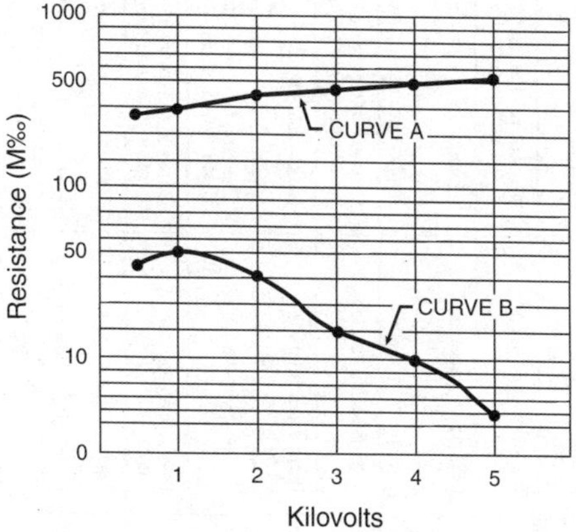

The resistance levels of good insulation, illustrated above by Curve A, remains approximately the same at different voltage levels. The resistance of cracked insulation (Curve B) decreases substantially at different voltage levels.

CONTACTOR AND MOTOR STARTER TROUBLESHOOTING GUIDE

Problem	Possible Cause	Corrective Action
Humming noise	Magnet pole faces misaligned	Realign, Replace magnet assembly if realignment is not possible.
	Too low voltage at coil	Measure voltage at coil. Check voltage rating of coil. Correct any voltage that is 10% less than coil rating.
	Pole face obstructed by foreign object, dirt, or rust	Remove any foreign object and clean as necessary. Never file pole faces.
Loud buzz noise	Shading coil broken	Replace coil assembly.
Controller fails to drop out	Voltage to coil not being removed	Measure voltage at coil. Trace voltage from coil to supply looking for shorted switch or contact if voltage is present.
	Worn or rusted parts causing binding	Clean rusted parts. Replace worn parts.
	Contact poles sticking	Checking for burning or sticky substance on contacts. Replace burned contacts. Clean dirty contacts.
	Mechanical interlock binding	Check to ensure interlocking mechanism is free to move when power is OFF. Replace faulty interlock.
Controller fails to pull in	No coil voltage	Measure voltage at coil terminals. Trace voltage loss from coil to supply voltage if voltage is not present.
	Too low voltage	Measure voltage at coil terminals. Correct voltage level if voltage is less than 10% of rated coil voltage. Check for a voltage drop as large loads are energized.
	Coil open	Measure voltage at coil. Remove coil if voltage is present and correct but coil does not pull in. Measure coil resistance for open circuit. Replace if open.

CONTACTOR AND MOTOR STARTER TROUBLESHOOTING GUIDE (cont.)

Problem	Possible Cause	Corrective Action
Controller fails to pull in (cont..)	Coil shorted	Shorted coil may show signs of burning. The fuse or breakers should trip if coil is shorted. Disconnect one side of coil and reset if tripped. Remove coil and check resistance for short if protection device does not trip. Replace shorted coil. Replace any coil that is burned.
	Mechanical obstruction	Remove any obstructions.
Contacts badly burned or welded	Too high inrush current	Measure inrush current. Check load for problem if higher-than-rated load current. Change to larger controller if load current is correct but excessive for controller.
	Too fast load cycling	Change to larger controller if load cycled ON and OFF repeatedly.
	Too large overcurrent protection device	Size overcurrent protection to load and controller.
	Short circuit	Check fuses or breakers. Clear any short circuit.
	Insufficient contact pressure	Check to ensure contacts are making good connection.
Nuisance tripping	Incorrect overload size	Check size of overload against rated load current. Size up if permissible per NEC®.
	Lack of temperature compensation	Correct setting of overload if controller and load are at different ambient temperatures.
	Loose connections	Check for loose terminal connection.

DIRECT CURRENT MOTOR TROUBLESHOOTING GUIDE

Problem	Possible Cause	Corrective Action
Motor will not start	Blown fuse or open CB	Test the OCPD. If voltage is present at the input, but not the output of the OCPD, the fuse is blown or the CB is open. Check the rating of the OCPD. It should be at least 125% of the motor's FLC.
	Motor overload on starter tripped	Allow overloads to cool. Reset overloads. If reset overloads do not start motor, test the starter.
	No brush contact	Check brushes. Replace, if worn.
	Open control circuit between incoming power and motor	Check for cleanliness, tightness, and breaks. Use a voltmeter to test the circuit starting with the incoming power and moving to the motor terminals. Voltage generally stops at the problem area.
Fuse, CB, or overloads retrip after service	Excessive load	If the motor is loaded to excess or is jammed, the circuit OCPD will open. Disconnect the load from the motor. If the motor now runs properly, check the load. If the motor does not run and the fuse or CB opens, the problem is with the motor or control circuit. Remove the motor from the control circuit and connect it directly to the power source. If the motor runs properly, the problem is in the control circuit. Check the control circuit. If the motor opens the fuse or CB again, the problem is in the motor. Replace or service the motor.
	Motor shaft does not turn	Disconnect the motor from the load. If the motor shaft still does not turn, the bearings are frozen. Replace or service the motor.
Brushes chip or break	Brush material is too weak or the wrong type for motor's duty rating	Replace with better grade or type of brush. Consult manufacturer if problem continues.
	Brush face is overheating and losing brush bonding material	Check for an overload on the motor. Reduce the load as required. Adjust brush holder arms.

DIRECT CURRENT MOTOR TROUBLESHOOTING GUIDE (cont.)

Problem	Possible Cause	Corrective Action
Brushes chip or break	Brush holder is too far from commutator	Too much space between the brush holder and the surface of the commutator allows the brush end to chip or break. Set correct space between brush holder and commutator.
	Brush tension is incorrect	Adjust brush tension so the brush rides freely on the commutator.
Brushes spark	Worn brushes	Replace worn brushes. Service the motor if rapid brush wear, excessive sparking, chipping, breaking, or chattering is present.
	Commutator is concentric	Grind commutator and undercut mica. Replace commutator if necessary.
	Excessive vibration	Balance armature. Check brushes. They should be riding freely.
Rapid brush wear	Wrong brush material, type, or grade	Replace with brushes recommended by manufacturer.
	Incorrect brush tension	Adjust brush tension so the brush rides freely on the commutator.
Motor overheats	Improper ventilation	Clean all ventilation openings, Vacuum or blow dirt out of motor with low-pressure, dry, compressed air.
	Motor is overloaded	Check the load for binding. Check shaft straightness. Measure motor current under operating conditions. If the current is above the listed current rating, remove the motor. Remeasure the current under no-load conditions. If the current is excessive under load but not when unloaded, check the load. If the motor draws excessive current when disconnected, replace or service the motor.

SHADED POLE MOTOR TROUBLESHOOTING GUIDE

Problem	Possible Cause	Corrective Action
Motor will not start	Blown fuse or open CB	Test OCPD. If voltage is present at the input, but not the output of the OCPD, the fuse is blown or the CB is open. Check the rating of the OCPD. It should be at least 125% of the motor's FLC.
	Motor overload on starter tripped	Allow overloads to cool. Reset overloads. If reset overloads do not start the motor, test the starter.
	Low or no voltage applied to motor	Check the voltage at the motor terminals. The voltage must be present and within 10% of the motor nameplate voltage. If voltage is present at the motor but the motor is not operating, remove the motor from the load the motor is driving. Reapply power to the motor. If the motor runs, the problem is with the load. If the motor does not run, the problem is with the motor. Replace or service the motor.
	Open control circuit between incoming power and motor	Check for cleanliness, tightness, and breaks. Use a voltmeter to test the circuit starting with the incoming power and moving to the motor terminals. Voltage generally stops at the problem area.
Fuse, CB, or overloads retrip after service	Excessive load	If the motor is loaded to excess or jammed, the circuit OCPD will open. Disconnect the load from the motor. If the motor now runs properly, check the load. If the motor does not run and the fuse or CB opens, the problem is with the motor or control circuit. Remove the motor from the control circuit and connect it directly to the power source. If the motor runs properly, the problem is in the control circuit. Check the control circuit. If the motor opens the fuse or CB again, the problem is in the motor. Replace or service the motor.
Excessive noise	Unbalanced motor or load	An unbalanced motor or load causes vibration, which causes noise. Realign the motor and load. Check for excessive end play or loose parts. If the shaft is bent, replace the rotor or motor.
	Dry or worn bearings	Dry or worn bearings cause noise. Bearings may be dry due to dirty oil, oil not reaching the shaft, or motor overheating. Oil bearings as recommended. If noise remains, replace the bearings or motor.
	Excessive grease	Ball bearings that have excessive grease may cause bearings to overheat. Overheated bearings cause noise. Remove excess grease.

SPLIT-PHASE MOTOR TROUBLESHOOTING GUIDE

Problem	Possible Cause	Corrective Action
Motor will not start	Thermal cutout switch is open	Reset the thermal switch. Caution: Resetting the thermal switch may automatically start the motor.
	Blown fuse or open CB	Test the OCPD. If voltage is present at the input, but not the output of the OCPD, the fuse is blown or the CB is open. Check the rating of the OCPD. It should be at least 125% of the motor's FLC.
	Motor overload on starter tripped	Allow overloads to cool. Reset overloads. If reset overloads do not start the motor, test the starter.
	Low or no voltage applied to motor	Check the voltage at the motor terminals. The voltage must be present and within 10% of the motor nameplate voltage. If voltage is present at the motor but the motor is not operating, remove the motor from the load the motor is driving. Reapply power to the motor. If the motor runs, the problem is with the load. If the motor does not run, the problem is with the motor. Replace or service the motor.
	Open control circuit between incoming power and motor	Check for cleanliness, tightness, and breaks. Use a voltmeter to test the circuit starting with the incoming power and moving to the motor terminals. Voltage generally stops at the problem area.
	Starting winding not receiving power	Check the centrifugal switch to make sure it connects the starting winding when the motor is OFF.
Fuse, CB, or overloads retrip after service	Blown fuse or open CB	Test the OCPD. If voltage is present at the input, but not the output of the OCPD, the fuse is blown or the CB is open. Check the rating of the OCPD. It should be at least 125% of the motor's FLC.
	Motor overload on starter tripped	Allow overloads to cool. Reset overloads. If reset overloads do not start the motor, test the starter.

SPLIT-PHASE MOTOR TROUBLESHOOTING GUIDE (cont.)

Problem	Possible Cause	Corrective Action
Fuse, CB, or overloads retrip after service	Low or no voltage applied to motor	Check the voltage at the motor terminals. The voltage must be present and within 10% of the motor nameplate voltage. If voltage is present at the motor but the motor is not operating, remove the motor from the load the motor is driving. Reapply power to the motor. If the motor runs, the problem is with the load. If the motor does not run, the problem is with the motor. Replace or service the motor.
	Open control circuit between incoming power and motor	Check for cleanliness, tightness, and breaks. Use a voltmeter to test the circuit starting with the incoming power and moving to the motor terminals. Voltage generally stops at the problem area.
	Motor shaft does not turn	Disconnect the motor from the load. If the motor shaft still does not turn, the bearings are frozen. Replace or service the motor.
Motor produces electric shock	Broken or disconnected ground strap	Connect or replace ground strap. Test for proper ground.
	Hot power lead at motor connecting terminals is touching motor frame	Disconnect the motor. Open the motor terminal box and check for poor connections, damaged insulation, or leads touching the frame. Service and test motor for ground.
	Motor winding shorted to frame	Remove, service, and test motor.
Motor overheats	Starting windings are not being removed from circuit as motor accelerates	When the motor is turned OFF, a distinct click should be heard as the centrifugal switch closes.
	Improper ventilation	Clean all ventilation openings. Vacuum or blow dirt out of motor with low-pressure, dry, compressed air.

SPLIT-PHASE MOTOR TROUBLESHOOTING GUIDE (cont.)

Problem	Possible Cause	Corrective Action
Motor overheats	Motor is overloaded	Check the load for binding. Check shaft straightness. Measure motor current under operating conditions. If current is above the listed current rating, remove the motor. Remeasure the current under no-load conditions. If the current is excessive under load but not when unloaded, check the load. If the motor draws excessive current when disconnected, replace or service the motor.
	Dry or worn bearings	Dry or worn bearings cause noise. The bearings may be dry due to dirty oil, oil not reaching the shaft, or motor overheating. Oil the bearings as recommended. If noise remains, replace the bearings or the motor.
	Dirty bearings	Clean or replace bearings.
Excessive noise	Excessive end-play	Check and play by trying to move the motor shaft in and out. Add end-play washers as required.
	Unbalanced motor or load	An unbalanced motor or load causes vibration, which causes noise. Realign the motor and load. Check for excessive end play or loose parts. If the shaft is bent, replace the rotor or motor.
	Dry or worn bearings	Dry or worn bearings cause noise. The bearings may be dry due to dirty oil, oil not reaching the shaft, or motor overheating. Oil the bearings as recommended. If noise remains, replace the bearings or the motor.
	Excessive grease	Ball bearings that have excessive grease may cause the bearings to overheat. Overheated bearings cause noise. Remove any excess grease.

THREE-PHASE MOTOR TROUBLESHOOTING GUIDE

Problem	Possible Cause	Corrective Action
Motor will not start	Wrong motor connections	Most 3ϕ motors are dual-voltage. Check for proper motor connections.
	Blown fuse or open CB	Test the OCPD. If voltage is present at the input, but not the output of the OCPD, the fuse is blown or the CB is open. Check the rating of the OCPD. It should be at least 125% of the motor's FLC.
	Motor overload on starter tripped	Allow overloads to cool. Reset overloads. If reset overloads do not start the motor, test the starter.
	Low or no voltage applied to motor	Check the voltage at the motor terminals. The voltage must be present and within 10% of the motor nameplate voltage. If voltage is present at the motor but the motor is not operating, remove the motor from the load the motor is driving. Reapply power to the motor. If the motor runs, the problem is with the load. If the motor does not run, the problem is with the motor. Replace or service the motor.
	Open control circuit between incoming power and motor	Check for cleanliness, tightness, and breaks. Use a voltmeter to test the circuit starting with the incoming power and moving to the motor terminals. Voltage generally stops at the problem area.
Fuse, CB, or overloads retrip after service	Power not applied to all three lines	Measure voltage at each power line. Correct any power supply problems.
	Blown fuse or open CB	Test the OCPD. If voltage is present at the input, but not the output of the OCPD, the fuse is blown or the CB is open. Check the rating of the OCPD. It should be at least 125% of the motor's FLC.
	Motor overload on starter tripped	Allow overloads to cool. Reset overloads. If reset overloads do not start the motor, test the starter.

THREE-PHASE MOTOR TROUBLESHOOTING GUIDE (cont.)

Problem	Possible Cause	Corrective Action
Fuse, CB, or overloads retrip after service	Low or no voltage applied to motor	Check the voltage at the motor terminals. The voltage must be present and within 10% of the motor nameplate voltage. If voltage is present at the motor but the motor is not operating, remove the motor from the load the motor is driving. Reapply power to the motor. If the motor runs, the problem is with the load. If the motor does not run, the problem is with the motor. Replace or service the motor.
	Open control circuit between incoming power and motor	Check for cleanliness, tightness, and breaks. Use a voltmeter to test the circuit starting with the incoming power and moving to the motor terminals. Voltage generally stops at the problem area.
	Motor shaft does not turn	Disconnect the motor from the load. If the motor shaft still does not turn, the bearings are frozen. Replace or service the motor.
Motor overheats	Motor is single phasing	Check each of the 3ϕ power lines for correct voltage.
	Improper ventilation	Clean all ventilation openings. Vacuum or blow dirt out of motor with low-pressure, dry, compressed air.
	Motor is overloaded	Check the load for binding. Check shaft straightness. Measure motor current under operating conditions. If the current is above the listed current rating, remove the motor. Remeasure the current under no-load conditions. If the current is excessive under load but not when unloaded, check the load. If the motor draws excessive current when disconnected, replace or service the motor.

MOTOR REPAIR AND SERVICE RECORD

Motor File #: ____________ Serial #: ____________

Date Installed: ____________ Motor Location: ____________

____

MFR: ____________ Type: ____________ Frame: ____________

HP: ____________ Volts: ____________ Amps: ____________

RPM: ____________ Filter Sizes: ____________

Date	Operation	Mechanic

SEMIANNUAL MOTOR MAINTENANCE CHECKLIST

Step	Operation	Mechanic
1	Turn OFF and lock out all power to the motor and its control circuit.	
2	Clean motor exterior and all ventilation ducts.	
3	Check motor's wire raceway.	
4	Check and lubricate bearings as needed.	
5	Check drive mechanism.	
6	Check brushes and commutator.	
7	Check slip rings.	
8	Check motor terminations.	
9	Check capacitors.	
10	Check all mounting bolts.	
11	Check and record line-to-line resistance.	
12	Check and record megohmmeter resistance from L1 to ground.	
13	Check motor controls.	
14	Reconnect motor and control circuit power supplies.	
15	Check line-to-line voltage for balance and level.	
16	Check line current draw against nameplate rating.	
17	Check and record inboard and outboard bearing temperatures.	

ANNUAL MOTOR MAINTENANCE CHECKLIST

Motor File #: ______________ Serial #: ______________

Date Installed: ______________ Motor Location: ______________

____

MFR: ______________ Type: ______________ Frame: ______________

HP: ______________ Volts: ______________ Amps: ______________

RPM: ______________ Date Serviced: ______________

Step	Operation	Mechanic
1	Turn OFF and lock out all power to the motor and its control circuit.	
2	Clean motor exterior and all ventilation ducts.	
3	Uncouple motor from load and disassemble.	
4	Clean inside of motor.	
5	Check centrifugal switch assemblies.	
6	Check rotors, armatures, and field windings.	
7	Check all peripheral equipment.	
8	Check bearings.	
9	Check brushes and commutator.	
10	Check slip rings.	
11	Reassemble motor and couple to load.	

ANNUAL MOTOR MAINTENANCE CHECKLIST (cont.)

Step	Operation	Mechanic
12	Flush old bearing lubricant and replace.	
13	Check motor's wire raceway.	
14	Check drive mechanism.	
15	Check motor terminations.	
16	Check capacitors.	
17	Check all mounting bolts.	
18	Check and record line-to-line resistance.	
19	Check and record megohmmeter resistance from T1 to ground.	
20	Check and record insulation polarization index.	
21	Check motor controls.	
22	Reconnect motor and control circuit power supplies.	
23	Check line-to-line voltage for balance and level.	
24	Check line current draw against nameplate rating.	
25	Check and record inboard and outboard bearing temperatures.	

SERVICE NOTES

CHAPTER 10
POWER TRANSMISSION

METHOD OF ALIGNING PULLEYS

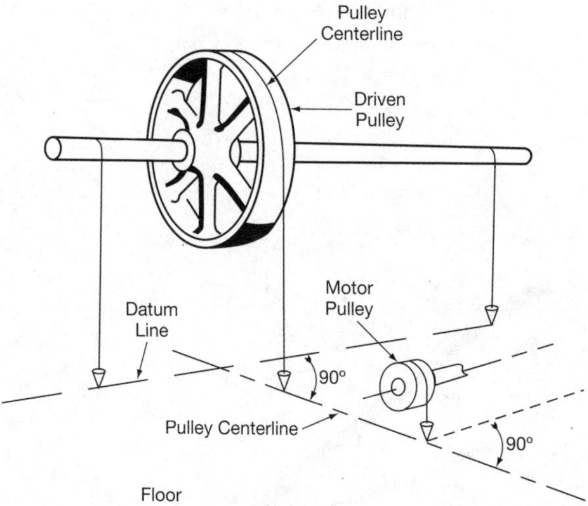

CALCULATING PULLEY DIAMETER

A pulley can be used to change the speed of a motor-driven device. To calculate the driven machine pulley diameter use the formula:

$$PD_m = \frac{PD_d \times N_d}{N_m}$$

PD_m = driven machine pulley diameter (in.)

PD_d = drive pulley diameter (in.)

N_d = motor drive speed (rpm)

N_m = driven machine speed (rpm)

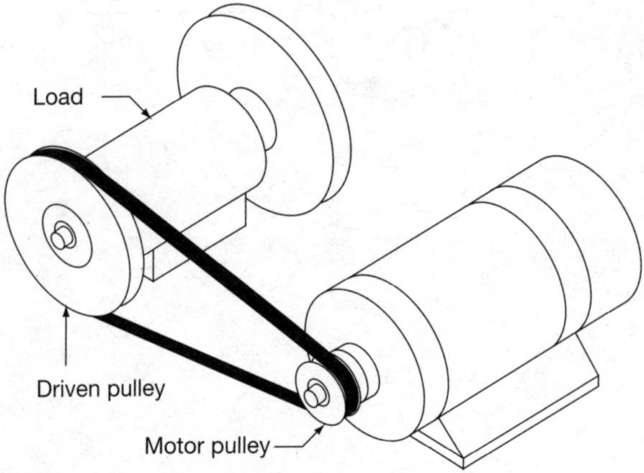

PULLEY AND GEAR CALCULATIONS

For single reduction or increase of speed by means of belting where the speed at which each shaft should run is known, and one pulley is in place:

Multiply the diameter of the pulley which you have by the number of revolutions per minute that its shaft makes; divide this product by the speed in revolutions per minute at which the second shaft should run. The result is the diameter of pulley to use.

Where both shafts with pulleys are in operation and the speed of one is known:

Multiply the speed of the shaft by diameter of its pulley and divide this product by diameter of pulley on the other shaft. The result is the speed of the second shaft.

Where a countershaft is used, to obtain size of main driving or driven pulley, or speed of main driving or driven shaft, it is necessary to calculate, as above, between the known end of the transmission and the countershaft, then repeat this calculation between the countershaft and the unknown end.

A set of gears of the same pitch transmits speeds in proportion to the number of teeth they contain. Count the number of teeth in the gear wheel and use this quantity instead of the diameter of pulley, mentioned above, to obtain number of teeth cut in unknown gear, or speed of second shaft.

FORMULAS FOR FINDING PULLEY SIZES

$$d = \frac{D \times S}{S'} \qquad D = \frac{d \times S'}{S}$$

d = diameter of driven pulley.

D = diameter of driving pulley.

S = number of revolutions per minute of driving pulley.

S' = number of revolutions per minute of driven pulley.

FORMULAS FOR FINDING GEAR SIZES

$$n = \frac{N \times S}{S'} \qquad N = \frac{n \times S'}{S}$$

n = number of teeth in pinion (driving gear).

N = number of teeth in gear (driven gear).

S = number of revolutions per minute of pinion.

S' = number of revolutions per minute of gear.

FORMULA TO DETERMINE SHAFT DIAMETER

The formula for determining the size of steel shaft for transmitting a given power at a given speed is as follows:

$$\text{diameter of shaft in inches} = \sqrt[3]{\frac{K \times HP}{RPM}}$$

when HP = the horsepower to be transmitted

RPM = speed of shaft

K = factor which varies from 50 to 125 depending on type of shaft and distance between supporting bearings.

For line shaft having bearings 8 feet apart:

K = 90 for turned shafting

K = 70 for cold-rolled shafting

FORMULA TO DETERMINE BELT LENGTH

The following formula is used to determine the length of belting:

$$\text{length of belt} = \frac{3.14\,(D + d)}{2} + 2\sqrt{X^2 + \left(\frac{D - d}{2}\right)^2}$$

when D = diameter of large pulley

d = diameter of small pulley

X = distance between centers of shafting

ADJUSTING BELT TENSION

A belt must be tight enough not to slip, but not so tight that it overloads the motor bearings. Belt tension is checked by placing a straightedge from pulley to pulley and measuring the amount of deflection at the midpoint. As a general rule, belt deflection should equal $\frac{1}{64}$" per inch of span. Belt tension is adjusted by moving the drive motor away from or closer to the driven device.

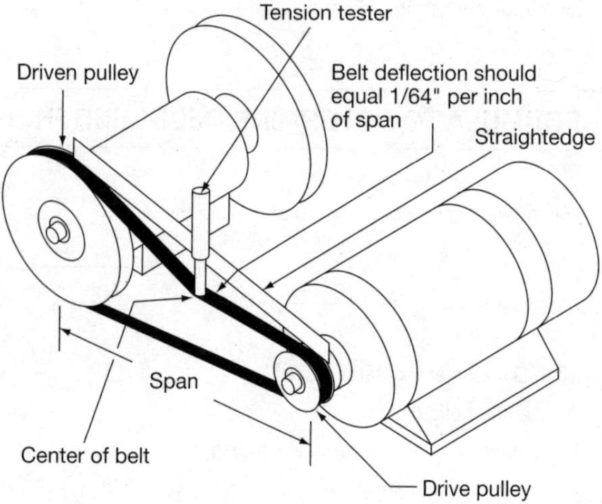

V-BELTS

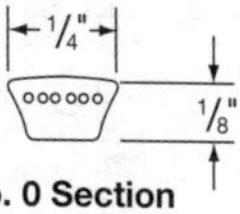

No. 0 Section
"2L"

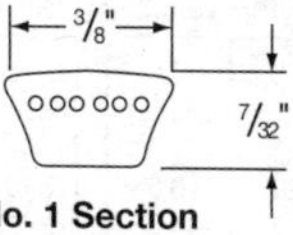

No. 1 Section
"3L"

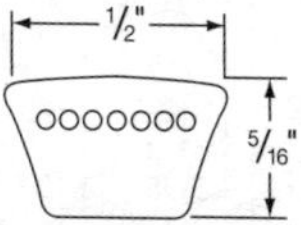

No. 2 Section
"4L"
A

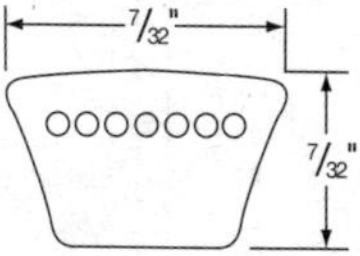

No. 3 Section
"5L"
B

V-BELT/MOTOR SIZE

3/8"
9.5 mm

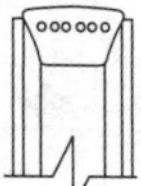

Up to .76 kW
1 HP

1/2"
11.7 mm

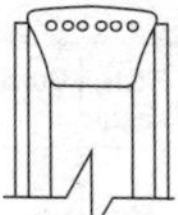

.56 to 4 kW
5 HP

21/32"
16.7 mm

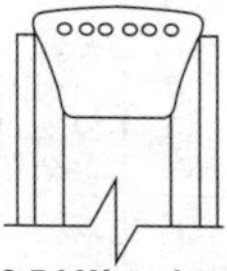

2.5 kW and up
3 HP

STANDARD "V" BELT LENGTHS

A BELTS

Standard Belt No.	Pitch Length	Outside Length
A26	27.3	28.0
A31	32.3	33.0
A35	36.3	37.0
A38	39.3	40.0
A42	43.3	44.0
A46	47.3	48.0
A51	52.3	53.0
A55	56.3	57.0
A60	61.3	62.0
A68	69.3	70.0
A75	76.3	77.0
A80	81.3	82.0
A85	86.3	87.0
A90	91.3	92.0
A96	97.3	98.0
A105	106.3	107.0
A112	113.3	114.0
A120	121.3	122.0
A128	129.3	130.0

B BELTS

Standard Belt No.	Pitch Length	Outside Length
B35	36.8	38.0
B38	39.8	41.0
B42	43.8	45.0
B46	47.8	49.0
B51	52.8	54.0
B55	56.8	58.0
B60	61.8	63.0
B68	69.8	71.0
B75	76.8	78.0
B81	82.8	84.0
B85	86.8	88.0
B90	91.8	93.0
B97	98.8	100.0
B105	106.8	108.0
B112	113.8	115.0
B120	121.8	123.0
B128	129.8	131.0
B136	137.8	139.0
B144	145.8	147.0
B158	159.8	161.0
B173	174.8	176.0
B180	181.8	183.0
B195	196.8	198.0
B210	211.8	213.0
B240	240.3	241.5
B270	270.3	271.5
B300	300.3	301.5

C BELTS

Standard Belt No.	Pitch Length	Outside Length
C51	53.9	55.0
C60	62.9	64.0
C68	70.9	81.0
C75	77.9	79.0
C81	83.9	85.0
C85	87.9	89.0
C90	92.9	94.0
C96	98.9	100.0
C105	107.9	109.0
C112	114.9	116.0
C120	122.9	124.0
C128	130.9	132.0
C136	138.9	140.0
C144	146.9	148.0
C158	160.9	162.0
C162	164.9	166.0
C173	175.9	177.0
C180	182.9	184.0
C195	197.9	199.0
C210	212.9	214.0
C240	240.9	242.0
C270	270.9	272.0
C300	300.9	302.0
C360	360.9	362.0
C390	390.9	392.0
C420	420.9	422.0

D BELTS

Standard Belt No.	Pitch Length	Outside Length
D120	123.3	125.0
D128	131.3	133.0
D144	147.3	149.0
D158	1613	163.0
D162	165.3	167.0
D173	176.3	178.0
D180	183.3	185.0
D195	198.3	200.0
D210	213.3	215.0
D240	240.8	242.0
D270	270.8	272.5
D300	300.8	302.5
D330	330.8	332.5
D360	360.8	362.5
D390	390.8	392.5
D420	420.8	422.5
D480	480.8	482.5
D540	540.8	542.5
D600	600.8	602.5

E BELTS

Standard Belt No.	Pitch Length	Outside Length
E180	184.5	187.5
E195	199.5	202.5
E210	214.5	217.5
E240	241.0	244.0
E270	271.0	274.0
E300	301.0	304.0
E330	331.0	334.0

E BELTS

Standard Belt No.	Pitch Length	Outside Length
E360	361.0	364.0
E390	391.0	394.0
E420	421.0	424.0
E480	481.0	484.0
E540	541.0	544.0
E600	601.0	604.0

STANDARD "V" BELT LENGTHS *(cont.)*

3V Belts		5V Belts		8V Belts	
3V250	25.0	5V500	50.0	8V1000	100.0
3V265	26.5	5V530	53.0	8V1060	106.0
3V280	28.0	5V560	56.0	8V1120	112.0
3V300	30.0	5V600	60.0	8V1180	118.0
3V315	31.5	5V630	63.0	8V1250	125.0
3V335	33.5	5V670	67.0	8V1320	132.0
3V355	35.5	5V710	71.0	8V1400	140.0
3V375	37.5	5V750	75.0	8V1500	150.0
3V400	40.0	5V800	80.0	8V1600	160.0
3V425	42.5	5V850	85.0	8V1700	170.0
3V450	45.0	5V900	90.0	8V1800	180.0
3V475	47.5	5V950	95.0	8V1900	190.0
3V500	50.0	5V1000	100.0	8V2000	200.0
3V530	53.0	5V1060	106.0	8V2120	212.0
3V560	56.0	5V1120	112.0	8V2240	224.0
3V600	60.0	5V1180	118.0	8V2360	236.0
3V630	63.0	5V1250	125.0	8V2500	250.0
3V670	67.0	5V1320	132.0	8V2650	265.0
3V710	71.0	5V1400	140.0	8V2800	280.0
3V750	75.0	5V1500	150.0	8V3000	300.0
3V800	80.0	5V1600	160.0	8V3150	315.0
3V850	85.0	5V1700	170.0	8V3350	335.0
3V900	90.0	5V1800	180.0	8V3550	355.0
3V950	95.0	5V1900	190.0	8V3750	375.0
3V1000	100.0	5V2000	200.0	8V4000	400.0
3V1060	106.0	5V2120	212.0	8V4250	425.0
3V1120	112.0	5V2240	224.0	8V4500	450.0
3V1180	118.0	5V2360	236.0	8V5000	500.0
3V1250	128.0	5V2500	250.0		
3V1320	132.0	5V2650	265.0		
3V1400	140.0	5V2800	280.0		
		5V3000	300.0		
		5V3150	315.0		
		5V3350	335.0		
		5V3550	355.0		

If the 60-inch "B" section belt shown is made 3/10 of an inch longer, it will be code marked 53 rather than 50. If made 3/10 shorter, it will be marked 47. While both have the belt number B60 they cannot be used in a set because of the difference in length.

TYPICAL CODE MARKING

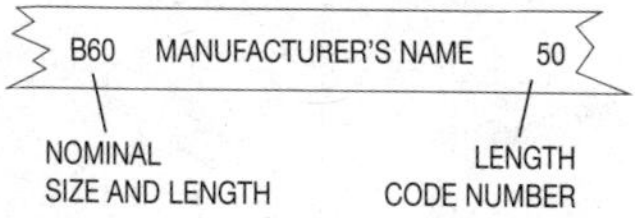

HORSEPOWER CAPACITIES OF LIGHT 4-PLY NYLON-STITCHED BELTS

Belt Speed (Feet Per Minute)	Pulley Diameter (inches)			
	1"	2"	3"	6"
1000	0.45	0.64	0.76	0.83
2000	0.94	1.27	1.38	1.56
3000	1.28	1.81	2.07	2.35
5000	1.89	2.73	3.18	3.75
6000	1.92	2.91	3.35	4.05
7000	1.93	2.94	3.57	4.51
8000	1.87	2.98	3.63	4.72
9000		2.92	3.68	4.79
10000			3.64	4.85
12000				4.81

HORSEPOWER CAPACITIES OF MEDIUM 4-PLY NYLON-STITCHED BELTS

Belt Speed (Feet Per Minute)	Pulley Diameter (inches)			
	1"	2"	3"	6"
1000	0.60	0.90	1.03	1.25
2000	1.15	1.67	1.95	2.33
3000	1.59	2.44	2.85	3.50
5000	1.80	3.09	3.80	4.85
6000	1.86	3.30	4.17	5.31
7000	1.82	3.35	4.25	5.54
8000		3.33	4.27	5.67
9000			4.21	5.71
10000				5.62

HORSEPOWER CAPACITIES OF HEAVY 4-PLY NYLON-STITCHED BELTS

Belt Speed (Feet Per Minute)	Pulley Diameter (inches)			
	1"	2"	3"	6"
1000	0.90	1.40	1.70	2.21
2000	1.66	2.06	3.21	4.19
3000	2.35	3.82	4.75	6.15
5000	3.03	5.51	7.00	9.49
6000	2.99	6.17	8.21	11.30
7000		6.10	8.39	12.20
8000			8.31	12.48
9000				12.25

HORSEPOWER CAPACITIES OF MEDIUM 4-PLY WOVEN ENDLESS COTTON BELTS

Belt Speed (Feet Per Minute)	Pulley Diameter (inches)			
	1-1½"	3"	4"	6"
500	0.25	0.35	0.40	0.50
1000	0.50	0.70	0.80	1.00
1500	0.80	1.00	1.30	1.50
2000	1.00	1.40	1.60	2.00
2500	1.30	1.80	2.00	2.30
3000	1.50	2.00	2.30	2.60
3500	1.80	2.40	2.50	3.00
4000	2.00	2.80	2.80	3.50
4500	2.10	3.20	3.20	4.10
5000	2.20	3.30	3.30	4.50

HORSEPOWER CAPACITIES PER INCH OF WIDTH OF REGULAR SINGLE-PLY DACRON BELTS

Belt Speed (Feet Per Minute)	Pulley Diameter (inches)						
	½"	1"	1½"	2"	3"	4"	6"
1000	0.5	1.0	1.0	1.2	1.4	1.5	1.6
2000	1.5	2.0	2.1	2.2	2.5	2.7	2.9
3000	1.9	2.4	3.0	3.3	3.9	4.1	4.4
4000	2.3	2.9	3.7	4.2	4.8	5.2	5.7
5000	2.5	3.2	4.2	4.7	5.5	6.0	6.7
6000	2.6	3.5	4.6	5.1	6.0	6.7	7.5
7000	2.6	3.6	4.8	5.4	6.5	7.4	8.4
8000	2.5	3.7	5.0	5.6	6.8	7.8	8.9
9000	2.3	3.8	5.1	5.7	6.9	8.0	9.2
10000	2.0	3.8	5.2	5.8	7.0	8.2	9.4
12000		3.6	5.2	5.8	7.0	8.2	9.5
15000		3.3	5.0	5.7	6.9	8.2	9.5
18000		3.0	4.8	5.6	6.9	8.1	9.5

HORSEPOWER CAPACITIES PER INCH OF WIDTH OF MEDIUM SINGLE-PLY DACRON BELTS

Belt Speed (Feet Per Minute)	Pulley Diameter (inches)						
	½"	1"	1½"	2"	3"	4"	6"
1000	0.33	0.528	0.66	0.79	0.924	0.99	1.056
2000	0.99	1.32	1.38	1.45	1.65	1.78	1.914
3000	1.25	1.58	1.98	2.17	2.594	2.70	2.90
4000	1.51	1.91	2.44	2.77	3.168	3.43	3.762
5000	1.65	2.11	2.77	3.10	3.63	3.96	4.42
6000	1.71	2.31	3.03	3.36	3.96	4.42	4.95
7000	1.71	2.37	3.16	3.56	4.29	4.884	5.54
8000	1.65	2.44	3.30	3.69	4.488	5.148	5.874
9000	1.51	2.50	3.36	3.76	4.554	5.28	6.072
10000	1.32	2.50	3.43	3.82	4.62	5.412	6.138
12000		2.37	3.43	3.82	4.62	5.412	6.20
15000		2.17	3.30	3.76	4.554	5.412	6.20
18000		1.98	3.16	3.69	4.554	5.346	6.27

HORSEPOWER CAPACITIES PER INCH OF WIDTH OF LIGHT SINGLE-PLY DACRON BELTS

Belt Speed (Feet Per Minute)	Pulley Diameter (inches)						
	½"	1"	1½"	2"	3"	4"	6"
1000	0.20	0.32	0.40	0.48	0.560	0.600	0.640
2000	0.60	0.80	0.84	0.88	1.00	1.08	1.16
3000	0.76	0.96	1.20	1.32	1.56	1.64	1.76
4000	0.92	1.16	1.48	1.68	1.92	2.08	2.28
5000	1.00	1.28	1.68	1.88	2.20	2.40	2.68
6000	1.04	1.40	1.84	2.04	2.40	2.68	3.00
7000	1.04	1.44	1.92	2.16	2.60	2.96	3.36
8000	1.00	1.48	2.00	2.24	2.72	3.12	3.56
9000	0.92	1.52	2.04	2.28	2.76	3.20	3.68
10000	0.80	1.52	2.08	2.32	2.80	3.28	3.72
12000		1.44	2.08	2.32	2.80	3.28	3.76
15000		1.32	2.00	2.28	2.76	3.28	3.76
18000		1.20	1.92	2.24	2.76	3.24	3.80

HORSEPOWER CAPACITIES OF ⅜" DIAMETER WOUND ENDLESS ROUND BELTS

Belt Speed (Feet Per Minute)	Pitch Diameter of Pulley (inches)					
	3"	3.4"	3.8"	4.2"	4.6"	5.0"
400	0.46	0.54	0.61	0.66	0.70	0.74
800	0.76	1.10	1.27	1.40	1.51	1.61
1200	1.02	1.27	1.46	1.62	1.76	1.87
1600	1.22	1.56	1.82	2.03	2.21	2.36
2000	1.30	1.80	2.13	2.40	2.62	2.80
2400	1.51	2.01	2.41	2.70	3.00	3.20
2800	1.59	2.20	2.61	3.00	3.30	3.55
3200	1.61	2.30	2.80	3.23	3.60	3.87
3600	1.60	2.34	3.00	3.54	3.98	4.13
4000	1.50	2.34	3.00	3.54	3.98	4.35
4800	1.16	2.16	2.95	3.60	4.03	4.57
5200	0.90	2.00	2.80	3.50	4.10	4.60
5600	0.54	1.71	2.64	3.38	4.00	4.50
6000	0.12	1.40	2.40	3.20	3.80	4.40

CORRECTION FACTORS FOR SMALL PULLEY ANGLES OF CONTACT LESS THAN 180°

Belt Type	45°	90°	110°	120°	130°	140°	150°	160°	170°
Nylon Stitched & Woven Endless			0.70	0.74	0.78	0.83	0.87	0.91	0.96
Single-Ply Dacron	0.30	0.60	0.70	0.74	0.78	0.83	0.87	0.91	0.96
Endless Round Belts		0.69	0.79	0.82	0.86	0.89	0.92	0.95	0.97

CORRECTION FACTORS FOR BELT VARIATIONS

Belt Type	Light 4-ply	Heavy 4-ply	Light 6-ply	Medium 6-ply	Heavy 6-ply	Medium 8-ply	Heavy 8-ply
Nylon Stitched			1.12	1.17		1.28	1.40
Woven Endless Cotton	0.33	1.50	0.50	1.50			

HORSEPOWER CAPACITIES OF ¼" DIAMETER BRAIDED ENDLESS ROUND BELTS

Belt Speed (Feet Per Minute)	Pulley Diameter, U-grooves (inches)				
	1½"	2"	3"	4"	5"
400	0.17	0.22	0.23	0.32	0.35
800	0.21	0.36	0.37	0.65	0.69
1200	0.33	0.45	0.45	0.79	0.83
1600	0.41	0.54	0.55	0.83	0.86
2000	0.44	0.58	0.59	0.98	1.04
2400	0.52	0.69	0.63	1.12	1.16
2800	0.54	0.72	0.68	1.24	1.28
3200	0.58	0.77	0.72	1.44	1.49
3600	0.63	0.84	0.74	1.59	1.60
4000	0.67	0.90	0.68	1.59	1.68
4800	0.69	0.92	0.59	1.57	1.70
5200	0.58	0.90	0.45	1.57	1.71
5600	0.63	0.85	0.28	1.56	1.70
6000	0.60	0.80	0.28	1.47	1.70

HORSEPOWER CAPACITIES OF ⅜" DIAMETER BRAIDED ENDLESS ROUND BELTS

Belt Speed (Feet Per Minute)	Pulley Diameter, U-grooves (inches)			
	2"	3"	4"	5"
500	0.45	0.50	0.70	0.78
750	0.72	0.80	1.44	1.61
1000	0.90	1.00	1.75	1.83
1250	1.08	1.20	1.85	1.90
1500	1.20	1.30	2.20	2.30
2000	1.25	1.40	2.50	2.58
2500	1.35	1.52	2.75	2.85
3000	1.45	1.61	3.20	3.30
3500	1.47	1.65	3.54	3.65
4000	1.37	1.52	3.54	3.72
4500	1.18	1.30	3.50	3.75
5000	0.91	1.00	3.50	3.77
5500	0.54	0.60	3.40	3.80
6000	0.27	0.30	3.30	3.75

HORSEPOWER CAPACITIES OF 9/16" DIAMETER WOUND ENDLESS ROUND BELTS

Belt Speed (Feet Per Minute)	Pitch Diameter of Pulley (inches)					
	5.0"	5.4"	5.8"	6.2"	6.6"	7.0"
400	0.94	1.00	1.10	1.20	1.22	1.25
800	1.60	1.80	1.92	2.10	2.17	2.25
1200	2.20	2.45	2.70	2.90	3.00	3.20
1600	2.70	3.00	3.30	3.70	3.80	4.00
2000	3.10	3.50	3.90	4.40	4.50	4.70
2400	3.50	4.00	4.40	4.95	5.10	5.40
2800	3.75	4.30	4.80	5.50	5.70	6.00
3200	3.95	4.61	5.20	5.90	6.10	6.50
3600	4.10	4.80	5.40	6.00	6.50	6.90
4000	4.10	4.90	5.60	6.20	6.80	7.20
4400	3.96	4.45	5.70	6.34	6.90	7.50
4800	3.75	4.30	5.60	6.40	7.00	7.60
5200	3.40	4.00	5.40	6.20	6.90	7.60
5600	2.96	3.60	5.10	6.00	6.70	7.40
6000	2.40	3.00	4.70	5.60	6.40	7.10

CHAPTER 11
COMMUNICATION SYSTEMS

TELEPHONE CIRCUIT OPERATION

RATING PAIRED COPPER BUILDING CABLES

- Fire Ratings . . . UL
  1 General Purpose . . . . Good
  2 Riser Grade . . . . Better
  3 Plenum Grade . . . . Best
- Performance Ratings
  - Cat 1 . . . . Not Rated . . . . Usually Residential
  - Cat 2 . . . . up to 1 MHz . . . . Not for Horizontal
  - Cat 3 . . . . up to 10 Mb/s . . . . Typical Voice/Data
  - Cat 4 . . . . up to 16 Mb/s . . . . Seldom Used
  - Cat 5 . . . . up to 100 Mb/s . . . . Emerging Applications

HORIZONTAL, FOUR-PAIR CAT 5 CABLING

- Characteristics
  - Twisted pair copper
  - 4 pairs
  - Rated to 100 MHz
  - Maximum length 259 feet from telecom closet to outlet.
  - Patch cords & cross-connects may add 20 feet.

TRADITIONAL TELEPHONE NETWORK (POTS)

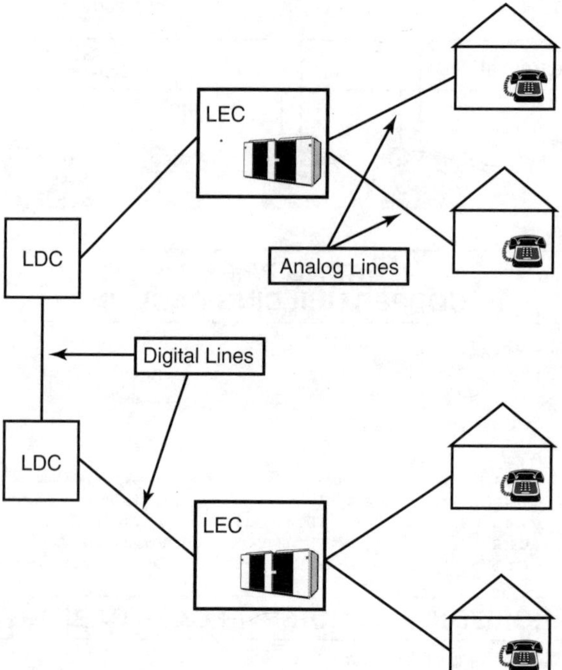

COMMON TELEPHONE CONNECTIONS

The most common and simplest type of communication installation is the single line telephone. The typical telephone cable (sometimes called quad cable) contains four wires, colored green, red, black, and yellow. A one line telephone requires only two wires to operate. In almost all circumstances, green and red are the two conductors used. In a common four-wire modular connector, the green and red conductors are found in the inside positions, with the black and yellow wires in the outer positions.

As long as the two center conductors of the jack (again, always green and red) are connected to live phone lines, the telephone should operate.

Two-line phones generally use the same four wire cables and jacks. In this case, however, the inside two wires (green and red) carry line 1, and the outside two wires (black and yellow) carry line 2.

COLOR CODING OF CABLES

The color coding of twisted-pair cable uses a color pattern that identifies not only what conductors make up a pair but also what pair in the sequence it is, relative to other pairs within a multipair sheath. This is also used to determine which conductor in a pair is the *tip* conductor and which is the *ring* conductor. (The tip conductor is the positive conductor, and the ring conductor is the negative conductor.)

The banding scheme uses two opposing colors to represent a single pair. One color is considered the primary while the other color is considered the secondary. For example, given the primary color of white and the secondary color of blue, a single twisted-pair would consist of one cable that is white with blue bands on it. The five primary colors are white, red, black, yellow, and violet.

In multi-pair cables the primary color is responsible for an entire group of pairs (five pairs total). For example, the first five pairs all have the primary color of white. Each of the secondary colors, blue, orange, green, brown, and slate, are paired in a banded fashion with white. This continues through the entire primary color scheme for all four primary colors (comprising 25 individual pairs). In larger cables (50 pairs and up), each 25-pair group is wrapped in a pair of ribbons, again representing the groups of primary colors matched with their respective secondary colors. These color coded band markings help cable technicians to quickly identify and properly terminate cable pairs.

EIA COLOR CODE

You should note that the new EIA color code calls for the following color coding:

Pair 1	–	White/Blue (white with blue stripe) and Blue
Pair 2	–	White/Orange and Orange
Pair 3	–	White/Green and Green
Pair 4	–	White/Brown and Brown

TWISTED-PAIR PLUGS AND JACKS

One of the more important factors regarding twisted-pair implementations is the cable jack or cross-connect block. These items are vital slnce without the proper interface, any twisted-pair cable would be relatively useless. In the twisted-pair arena, there are three major types of twisted-pair jacks:

RJ-type connectors (phone plugs)

Pin-connector

Genderless connectors (IBM sexless data connectors)

The RJ-type (registered jack) name generally refers to the standard format used for most telephone jacks. The term pin-connector refers to twisted pair connectors, such as the RS-232 connector, which provide connection through male and female pin receptacles. Genderless connectors are connectors in which there is no separate male or female component; each component can plug into any other similar component.

STANDARD PHONE JACKS

The standard phone jack is specified by a variety of different names, such as RJ and RG, which refer to their physical and electrical characteristics. These jacks consist of a male and a female component. The male component snaps into the female receptacle. The important point to note, however, is the number of conductors each type of jack can support.

Common configurations for phone jacks include support for four, six, or eight conductors. A typical example of a four-conductor jack, supporting two twisted-pairs, would be the one used for connecting most telephone handsets to their receivers.

A common six-conductor jack, supporting three twisted-pairs, is the RJ-11 jack used to connect most telephones to the telephone company or PBX systems. An example of an eight-conductor jack Is the R-45 jack, which is intended for use under the ISDN system as the user-site interface for ISDN terminals.

For building wiring, the six-conductor and eight-conductor jacks are popular, with the eight-conductor jack increasing in popularity, as more corporations install twisted-pair in four-pair bundles for both voice and data. The eight-conductor jack, in addition to being used for ISDN, is also specified by several other popular applications, such as the new IEEE 802.3 10 BaseT standard for Ethernet over twisted-pair.

These types of jacks are often keyed, so that the wrong type of plug cannot be inserted into the jack. There are two kinds of keying -- side keying, and shift keying.

Side keying uses a piece of plastic that is extended to one side of the jack. This type is often used when multiple jacks are present.

Shift keying entails shifting the position of the snap connector to the left or right of the jack, rather than leaving it in its usual center position. Shift keying Is more commonly used for data connectors than for voice connectors.

Note that while we say that these jacks are used for certain types of systems (data, voice, etc.) this is not any type of standard. They can be used as you please.

PIN CONNECTORS

There are any number of pin type connectors available. The most familiar type is the RS-232 jack that is commonly used for computer ports. Another popular type of pin connector is the DB type connector, which is the round connector that is commonly used for computer keyboards.

The various types of pin connectors can be used for terminating as few as five (the DB type), or more than 50 (the RS type) conductors.

50-pin *champ* type connectors are often used with twisted-pair cables, when connecting to cross-connect equipment, patch panels, and communications equipment such as Is used for networking.

CROSS CONNECTIONS

Cross connections are made at terminal *blocks.* A block is typically a rectangular, white plastic unit, with metal connection points. The most common type is called a punchdown block. This is the kind that you see on the back wall of a business, where the main telephone connections are made.The wire connections are made by pushing the insulated wires into their places. When "punched" down, the connector cuts through the insulation, and makes the appropriate connection.

Connections are made between punch-down blocks by using *patch cords*, which are short lengths of cable that can be terminated into the punch-down slots, or that are equipped with connectors on each end.

When different systems must be connected together, cross-connects are used.

CATEGORY CABLING

Category 1 cable is the old standard type of telephone cable, with four conductors colored green, red, black, and yellow. Also called quad cable.

Category 2 was an old IBM cabling system. It is almost never used for modern communications.

Category 3 cable is used for digital voice and data transmlsslon rates up to 10 Mbps (Megabits per second). Common types of data transmission over this communications cable would be UTP Token Ring (4 Mbps) and 10Base-T (10 Mbps).

Category 4 cable is used for voice and data transmission rates up to 16 Mbps. A typical type of transmission over this system would be UTP Token Ring (16 Mbps).

Category 5 cable is used for sending voice and data at speeds up to 100Mbit/s (megabits per second).

This includes signals used under the FDDI communications standard.

INSTALLATION REQUIREMENTS

Article 800 of the NEC covers communication circuits, such as telephone systems and outside wiring for fire and burglar alarm systems. Generally these clrcults must be separated from power circuits and grounded. In addition, all such circuits that run out of doors (even if only partially) must be provided with circuit protectors (surge or voltage supressors).

The requirements for these installations are as follows:

CONDUCTORS ENTERING BUILDINGS

If communications and power conductors are supported by the same pole, or run parallel in span, the following conditions must be met:

1. Wherever possible, communications conductors should be located below power conductors.
2. Communications conductors cannot be connected to crossarms.
3. Power service drops must be separated from communications service drops by at least 12 inches.

Above roofs, communications conductors must have the following clearances:

1. Flat roofs: 8 feet.
2. Garages and other auxiliary buildings: None required.
3. Overhangs, where no more than 4 feet of communications cable will run over the area: 18 inches.
4. Where the roof slope is 4 inches rise for every 12 inches horizontally: 3 feet.

Underground communications conductors must be separated from power conductors in manhole or handholes by brick, concrete, or tile partitions.

Communications conductors should be kept at least 6 feet away from lightning protection system conductors.

CIRCUIT PROTECTION

Protectors are surge arresters designed for the specific requirements of communications circuits. They are required for all aerial circuits not confined with a *block*. (Block here means city block.) They must be installed on all clrcults with a block that could accidentally contact power circuits over 300 volts to ground. They must also be listed for the type of installation.

Other requirements are the following:

Metal sheaths of any communications cables must be grounded or interrupted with an insulating joint as close as practicable to the point where they enter any building (such point of entrance being the place where the communications cable emerges through an exterior wall or concrete floor slab, or from a grounded rigid or intermediate metal conduit).

Grounding conductors for communications circuits must be copper or some other corrosion-resistant material, and have insulation suitable for the area in which it is installed.

Communications grounding conductors may be no smaller than No. 14.

The grounding conductor must be run as directly as possible to the grounding electrode, and be protected if necessary.

If the grounding conductor Is protected by metal raceway, it must be bonded to the grounding conductor on both ends.

CIRCUIT PROTECTION *(cont'd)*

Grounding electrodes for communications ground may be any of the following:
1. The grounding electrode of an electrical power system.
2. A grounded interior metal piping system (Avoid gas piping systems for obvious reasons.)
3. Metal power service raceway.
4. Power service equipment enclosures.
5. A separate grounding electrode.

If the building being served has no grounding electrode system, the following can be used as a grounding electrode:
1. Any acceptable power system grounding electrode. (See Section 250-81.)
2. A grounded metal structure.
3. A ground rod or pipe at least 5 feet long and 1/2 inch in diameter. This rod should be driven into damp (if possible) earth, and kept separate from any lightning protection system grounds or conductors.

Connections to grounding electrodes must be made with approved means.

If the power and communications systems use separate grounding electrodes, they must be bonded together with a No. 6 copper conductor. Other electrodes may be bonded also. This is not required for mobile homes.

For mobile homes, if there is no servlce equipment or disconnect within 30 feet of the mobile home wall, the communications circuit must have its own grounding electrode. In this case, or if the mobile home is connected with cord and plug, the communications circuit protector must be bonded to the mobile home frame or grounding terminal with a copper conductor no smaller than No. 12.

INTERIOR COMMUNICATIONS CONDUCTORS

Communications conductors must be kept at least 2 inches away from power or Class 1 conductors, unless they are permanently separated from them or unless the power or Class 1 conductors are enclosed in one of the following:
1. Raceway.
2. Type AC, MC, UF, NM, or NM cable, or metal-sheathed cable.

Communications cables are allowed in the same raceway, box, or cable with any of the following:
1. Class 2 and 3 remote-control, signaling, and power-limited circuits.
2. Power-limited fire protective signaling systems.
3. Conductive or nonconductive optical fiber cables.
4. Community antenna television and radio distribution systems.

Communications conductors are not allowed to be in the same raceway or fitting with power or Class 1 circuits.

Communications conductors are not allowed to be supported by raceways unless the raceway runs directly to the piece of equipment the communications circuit serves.

Openings through fire-resistant floors, walls, etc. must be sealed with an appropriate firestopping material.

Any communications cables used in plenums or environmental air-handling spaces must be listed for such use.

STANDARD TELECOM COLOR CODING

PAIR #	TIP (+) COLOR	RING (–) COLOR
1	White	Blue
2	White	Orange
3	White	Green
4	White	Brown
5	White	Slate
6	Red	Blue
7	Red	Orange
8	Red	Green
9	Red	Brown
10	Red	Slate
11	Black	Blue
12	Black	Orange
13	Black	Green
14	Black	Brown
15	Black	Slate
16	Yellow	Blue
17	Yellow	Orange
18	Yellow	Green
19	Yellow	Brown
20	Yellow	Slate
21	Violet	Blue
22	Violet	Orange
23	Violet	Green
24	Violet	Brown
25	Violet	Slate

MODULAR JACK STYLES

8-Position | **8-Position Keyed** | **6-Position** | **6-Position Modified**

There are four basic modular jack styles. The 8-position and 8-position keyed modular jacks are commonly and incorrectly referred to as RJ45 and keyed RJ45 (respectively). The 6-position modular jack is commonly referred to as RJ11. Using these terms can sometimes lead to confusion since the RJ designations actually refer to very specific wiring configurations called Universal Service Ordering Codes (USOC). The designation 'RJ' means Registered Jack. Each of these 3 basic jack styles can be wired for different RJ configurations. For example, the 6-position jack can be wired as a RJ11C (1-Pair), RJ14C (2-Pair), or RJ25C (3-Pair) configuration. An 8-position jack can be wired for configurations such as RJ61C (4-Pair) and RJ48C. The keyed 8-position jack can be wired for RJ45S, RJ46S and RJ47S. The fourth modular jack style is a modified version of the 6-position jack (modified modular jack or MMJ). It was designed by DEC along with the modified modular plug (MMP) to eliminate the possibility of connecting DEC data equipment to voice lines and vice versa.

COMMON WIRING CONFIGURATIONS

The TIA and AT&T wiring schemes are the two that have been adopted by EIA/TIA-568. They are nearly identical except that pairs two and three are reversed. TIA is the preferred scheme because it is compatible with 1 or 2-pair USOC Systems. Either configuration can be used for Integrated Services Digital Network (ISDN) applications.

TIA (T568A)

Pair ID	PIN #
T1	5
R1	4
T2	3
R2	6
T3	1
R3	2
T4	7
R4	8

AT&T (T568B)

Pair ID	PIN #
T1	5
R1	4
T2	1
R2	2
T3	3
R3	6
T4	7
R4	8

COMMON WIRING CONFIGURATIONS *(cont'd)*

USOC wiring is available for 1-, 2-, 3-, or 4-pair systems. Pair 1 occupies the center conductors, pair 2 occupies the next two contacts out, etc. One advantage to this scheme is that a 6-position plug configured with 1, 2, or 3 pairs can be inserted into an 8-position jack and maintain pair continuity; a note of warning though, pins 1 and 8 on the jack may become damaged from this practice. A disadvantage is the poor transmission performance associated with this type of pair sequence.

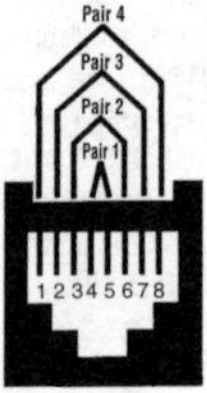

USOC
4-Pair

Pair ID	PIN #
T1	5
R1	4
T2	3
R2	6
T3	2
R3	7
T4	1
R4	8

USOC
1-, 2-, or 3-Pair

Pair ID	PIN #
T1	4
R1	3
T2	2
R2	5
T3	1
R3	6

ETHERNET 10 BASE-T

Ethernet 10BASE-T wiring specifies an 8-position jack but uses only two pairs. These are pairs two and three of TIA schemes.

Pair ID	PIN #
T1	1
R1	2
T2	3
R2	6

TELEPHONE CONNECTIONS

Typical Inside Wire

Type of Wire	Pair No.	Pair Color Matches	
2-pair Wire	1 2	Green Black	Red Yellow
3-pair Wire	1 2 3	White/Blue White/Orange White/Green	Blue/White Orange/White Green/White

Inside Wire Connecting Terminations

Wire Color		Wire Function	
2-pair wire	3-pair wire	Service w/o Dial Light	Service with Dial Light
Green Red Black Yellow	White/Blue Blue/White White/Orange Orange/White	Tip Ring Not Used Ground	Tip Ring Transformer Transformer

Typical Fasteners and Recommended Spacing Distances

Fasteners	Horizontal	Vertical	From Corner
Wire clamp	16 in.	16 in.	2 in.
Staples(wire)	7.5 in.	7.5 in.	2 in.
Bridle Rings*	4 ft.		2–8.5 in.*
Drive Rings**	4 ft.	8 ft.	2–8.5 in.*

*When changing direction the fasteners should be spaced to hold the wire at approximately a 45-degree angle.

**To avoid possible injury do not use drive rings below a 6 foot clearance level; instead, use bridle rings.

SEPARATION AND PHYSICAL PROTECTION FOR PREMISES WIRING

This table applies only to telephone wiring from the Network Interface or other telephone company-provided modular jacks to telephone equipment. Minimum separations between telephone wiring whether located inside or attached to the outside of buildings and other types of wiring involved are as follows. Separations apply to crossing and to parallel runs (minimum separations).

Types of Wire Involved		Minimum Separations	Wire Crossing Alternatives
Electric Supply	Bare light or power wire of any voltage	5 ft.	No Alternative
	Open wiring not over 300 volts	2 in.	See Note 1.
	Wires in conduit or in armored or nonmetallic sheath cable, or power ground wires	None	N/A
Radio & TV	Antenna lead-in and ground wires	4 in.	See Note 1.
Signal or Control Wires	Open wiring or wires in conduit or cable	None	N/A
Comm. Wires	Community Television systems coaxial cables with grounded shielding	None	N/A
Telephone Drop Wire	Using fused protectors Using fuseless protector or where no protector wiring from transformer	2 in.	See Note 1. None
Sign	Neon Signs and associated wiring from transformer		6 in. No Alternative
Lightning Systems	Lightning rods and wires		6 ft.

NOTE 1: If minimum separations cannot be obtained, additional protection of a plastic tube, wire guard, or two layers of vinyl tape extending 2 inches beyond each side of object being crossed must be provided.

ETHERNET 10BASE-T STRAIGHT THRU PATCH CORD

RJ45 Plug			RJ45 Plug
T2	1	White/Orange	1 TxData + pair 2/—
R2	2	Orange	2 TxData −
T3	3	White/Green	3 RecvData +
R1	4	Blue	4 pair3
T1	5	White/Blue	5
R3	6	Green	6 RecvData −
T4	7	White/Brown	7
R4	8	Brown	8

ETHERNET 10BASE-T CROSSOVER PATCH CORD

- This cable is used to cascade hubs, or for connecting two Ethernet stations back-to-back without a hub. Note pin numbering of straight-thru patch cord.

RJ45 Plug	RJ45 Plug
1 Tx+	Rx+ 3
2 TX	Rx− 6
3 RX+	Tx+ 1
6 Rx+	Tx− 2

DIGITAL PATCH CABLE (DPC) CODING

Pair 1: Green & Red

Pair 2: Yellow & Black

Pair 3: Blue & Orange

Pair 4: Brown & Gray

DISTANCE LIMITS FOR HORIZONTAL CABLING

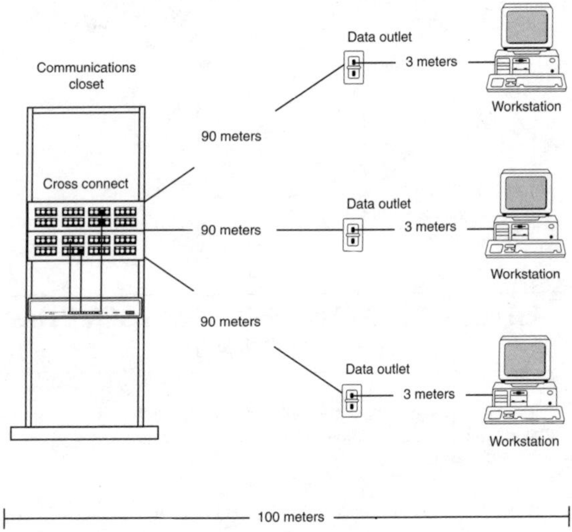

In addition to 90 meters of horizontal cable, 10 meters is allowed for work area and communications closet cables.

STANDARD MATERIALS FOR STRUCTURED CABLING (568) SYSTEMS

Four-pair 100 ohm UTP cables. The cable consists of 24 AWG thermoplastic insulated conductors formed into four individually twisted pairs and enclosed by a thermoplastic jacket. Four-pair, 22 AWG cables which meet the transmission requirements may also be used. Four-pair, *shielded* twisted pair cables which meet the transmission requirements may also be used.

The diameter over the insulation shall be 1.22 mm (0.048 in) max.

The pair twists of any pair shall not be exactly the same as any other pair. The pair twist lengths shall be selected by the manufacturer to assure compliance with the crosstalk requirements of this standard.

Color Codes.

Pair 1	White-Blue (W-BL)	Blue (BL)
Pair 2	White-Orange (W-O)	Orange (O)
Pair 3	White-Green (W-G)	Green (G)
Pair 4	White-Brown (W-BR)	Brown (BR)

ETHERNET FAILURES

Ethernets fail in three common ways:

1. A nail or other object can breaking one of the conductors.
2. A screw or other object can touch one or more of the conductors and short them to an external grounded metal shield, conduit, or other grounded metal.
3. A station on the network can break down and start to generate a continuous stream of electronic noise, thus blocking legitimate transmissions.

COMPUTER CIRCUITS

A network is a collection of electrical signaling circuits, each carrying digital signals between pieces of equipment. There are power sources, conductors, and loads involved in the process. The **power source** is a network device that transmits an electrical signal. **The conductors** are the wires that the signal travels over to reach its destination (another network device). The receiver is **the load.** These items, connected together, make up a complete circuit.

In the computer world, the electric signal transmitted by an energy source is a digital signal known as a **pulse.** Pulses are simply the presence of voltage and a lack of the presence of voltage, generated in a sequence. These pulses are used to represent a series of ones and zeros and ones (the presence of voltage being a 1, and the absence of voltage being a 0). These zeros or ones are called *bits.* Many years ago, computer engineers began using groupings of eight bits to represent digital "words," and to this day, a series of 8 bits is called a *byte.* These terms are used everywhere in the computer fields.

The key to successful signal transmission is that when a load receives an electrical signal, the signal must have a voltage level and configuration consistent with what had been originally transmitted by the energy source. If the signal has undergone too much corruption, the load won't be able to interpret it accurately.

A good cable will transfer a signal without too much distortion of the signal, while a bad cable will render a signal useless.

DATA SIGNAL TRANSMISSION

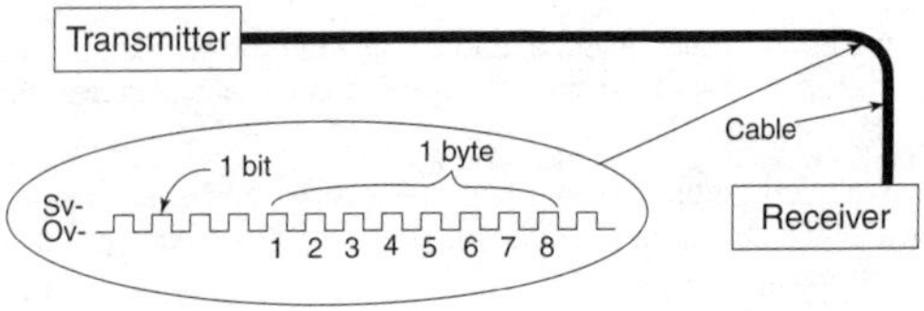

CABLE COLORS

Horizontal voice cables	Blue
Inter-building backbone	Brown
Second-level backbone	Gray
Network connections & auxiliary circuits	Green
Demarcation point, telephone cable from Central Office	Orange
First-level backbone	Purple
Key-type telephone systems	Red
Horizontal data cables, computer & PBX equipment	Silver or White
Auxiliary, maintenance & security alarms	Yellow

SEPARATION FROM SOURCES OF INTERFERENCE

Unshielded data cables should not be installed near sources of electromagnetism. There is a standard that specifies these distances for structured data cabling systems. EIA/TIA-569, the cabling pathways standard, specifies the following:

Minimum Separation Distance from Power Source at 480V or less

CONDITION	<2kVA	2-5kVA	>5kVA
Unshielded power lines or electrical equipment in proximity to open or non-metal pathways	5 in.	12 in.	24 in.
Unshielded power lines or electrical equipment in proximity to grounded metal conduit pathway	2.5 in.	6 in.	12 in.
Power lines enclosed in a grounded metal conduit (or equivalent shielding (in proximity to grounded metal conduit pathway	—	6 in.	12 in.
Transformers & electric motors	40 in.	40 in.	40 in.
Fluorescent lighting	12 in.	12 in.	12 in.

MINIMUM BENDING RADII

According to a draft version of EIA-568, the minimum bend radius for UTP is 4 times outside cable diameter, or about one inch. For multi-pair cables the minimum bending radius is 10 × outside diameter. The minimum bend radii for Type 1A Shielded Twisted Pair (100 Mb/s STP) is 7.5 cm (3-in) for non-plenum cable, 15 cm (6-in) for the stiffer plenum-rated kind.

For optical cables not under tension, the minimum bend radius is 10 times diameter; and for cables under tension, no less than 20 times cable diameter. The standard goes on to state that no optical cable will be bent on a radius less than 3.0 cm (1.18-in).

A different standard, ISO DIS 11801 (essentially a parallel standard to the one mentioned above), for 100 ohm and 120 ohm balanced cable lists three different minimum bend radii. Minimum for pulling during installation is 8 times cable diameter, minimum installed radius is 6 times for riser cable, and 4 times cable diameter for horizontal runs. For fiber optic cables, the requirements are the same as those stated above.

Some manufacturers recommendations differ from the above, so it is worth checking the spec sheet for the cable you plan to use.

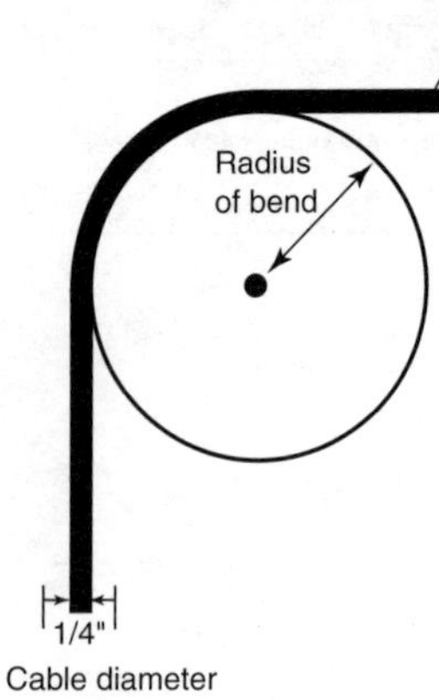

Radius of 10 × diameter = $2^1/_2$"

Radius of 20 × diameter = 5"

CHAPTER 12
CONVERSION FACTORS, MEASUREMENTS, TOOLS & MATERIALS

COMMONLY USED CONVERSION FACTORS

Multiply	By	To Obtain
Acres	43,560	Square feet
Acres	1.562×10^{-3}	Square miles
Acre-Feet	43,560	Cubic feet
Amperes per sq. cm.	6.452	Amperes per sq. in.
Amperes per sq. in.	0.1550	Amperes per sq. cm.
Ampere-Turns	1.257	Gilberts
Ampere-Turns per cm.	2.540	Ampere-turns per in.
Ampere-Turns per in.	0.3937	Ampere-turns per cm.
Atmospheres	76.0	Cm. of mercury
Atmospheres	29.92	Inches of mercury
Atmospheres	33.90	Feet of water
Atmospheres	14.70	Pounds per sq. in.
British thermal units	252.0	Calories
British thermal units	778.2	Foot-pounds
British thermal units	3.960×10^{-4}	Horsepower-hours
British thermal units	0.2520	Kilogram-calories
British thermal units	107.6	Kilogram-meters
British thermal units	2.931×10^{-4}	Kilowatt-hours
British thermal units	1,055	Watt-seconds
B.t.u. per hour	2.931×10^{-4}	Kilowatts
B.t.u. per minute	2.359×10^{-2}	Horsepower
B.t.u. per minute	1.759×10^{-2}	Kilowatts
Bushels	1.244	Cubic feet
Centimeters	0.3937	Inches
Circular mils	5.067×10^{-6}	Square centimeters
Circular mils	0.7854×10^{-6}	Square inches

COMMONLY USED CONVERSION FACTORS (cont.)

Multiply	By	To Obtain
Circular mils	0.7854	Square mils
Cords	128	Cubic feet
Cubic centimeters	6.102×10^{-6}	Cubic inches
Cubic feet	0.02832	Cubic meters
Cubic feet	7.481	Gallons
Cubic feet	28.32	Liters
Cubic inches	16.39	Cubic centimeters
Cubic meters	35.31	Cubic feet
Cubic meters	1.308	Cubic yards
Cubic yards	0.7646	Cubic meters
Degrees (angle)	0.01745	Radians
Dynes	2.248×10^{-6}	Pounds
Ergs	1	Dyne-centimeters
Ergs	7.37×10^{-6}	Foot-pounds
Ergs	10^{-7}	Joules
Farads	10^{6}	Microfarads
Fathoms	6	Feet
Feet	30.48	Centimeters
Feet of water	.08826	Inches of mercury
Feet of water	304.8	Kg. per square meter
Feet of water	62.43	Pounds per square ft.
Feet of water	0.4335	Pounds per square in.
Foot-pounds	1.285×10^{-2}	British thermal units
Foot-pounds	5.050×10^{-7}	Horsepower-hours
Foot-pounds	1.356	Joules
Foot-pounds	0.1383	Kilogram-meters
Foot-pounds	3.766×10^{-7}	Kilowatt-hours
Gallons	0.1337	Cubic feet
Gallons	231	Cubic inches
Gallons	3.785×10^{-3}	Cubic meters
Gallons	3.785	Liters
Gallons per minute	2.228×10^{-3}	Cubic feet per sec.
Gausses	6.452	Lines per square in.
Gilberts	0.7958	Ampere-turns
Henries	10^{3}	Millihenries
Horsepower	42.41	B.t.u. per min.
Horsepower	2,544	B.t.u. per hour

COMMONLY USED CONVERSION FACTORS (cont.)

Multiply	By	To Obtain
Horsepower	550	Foot-pounds per sec.
Horsepower	33,000	Foot-pounds per min.
Horsepower	1.014	Horsepower (metric)
Horsepower	10.70	Kg. calories per min.
Horsepower	0.7457	Kilowatts
Horsepower (boiler)	33,520	B.t.u. per hour
Horsepower-hours	2,544	British thermal units
Horsepower-hours	1.98×10^6	Foot-pounds
Horsepower-hours	2.737×10^5	Kilogram-meters
Horsepower-hours	0.7457	Kilowatt-hours
Inches	2.540	Centimeters
Inches of mercury	1.133	Feet of water
Inches of mercury	70.73	Pounds per square ft.
Inches of mercury	0.4912	Pounds per square in.
Inches of water	25.40	Kg. per square meter
Inches of water	0.5781	Ounces per square in.
Inches of water	5.204	Pounds per square ft.
Joules	9.478×10^{-4}	British thermal units
Joules	0.2388	Calories
Joules	10^7	Ergs
Joules	0.7376	Foot-pounds
Joules	2.778×10^{-7}	Kilowatt-hours
Joules	0.1020	Kilogram-meters
Joules	1	Watt-seconds
Kilograms	2.205	Pounds
Kilogram-calories	3.968	British thermal units
Kilogram meters	7.233	Foot-pounds
Kg. per square meter	3.281×10^{-3}	Feet of water
Kg. per square meter	0.2048	Pounds per square ft.
Kg. per square meter	1.422×10^{-3}	Pounds per square in.
Kilolines	10^3	Maxwells
Kilometers	3.281	Feet
Kilometers	0.6214	Miles
Kilowatts	56.87	B.t.u. per min.
Kilowatts	737.6	Foot-pounds per sec.
Kilowatts	1.341	Horsepower
Kilowatts-hours	3409.5	British thermal units

COMMONLY USED CONVERSION FACTORS (cont.)

Multiply	By	To Obtain
Kilowatts-hours	2.655×10^6	Foot-pounds
Knots	1.152	Miles
Liters	0.03531	Cubic feet
Liters	61.02	Cubic inches
Liters	0.2642	Gallons
Log N_e or in N	0.4343	Log_{10} N
Log N	2.303	Log_e N or in N
Lumens per square ft.	1	Foot-candles
Maxwells	10^{-3}	Kilolines
Megalines	10^6	Maxwells
Megaohms	10^6	Ohms
Meters	3.281	Feet
Meters	39.37	Inches
Meter-kilograms	7.233	Pound-feet
Microfarads	10^{-6}	Farads
Microhms	10^{-6}	Ohms
Microhms per cm. cube	0.3937	Microhms per in. cube
Microhms per cm. cube	6.015	Ohms per mil. foot
Miles	5,280	Feet
Miles	1.609	Kilometers
Miner's inches	1.5	Cubic feet per min.
Ohms	10^{-6}	Megohms
Ohms	10^6	Microhms
Ohms per mil foot	0.1662	Microhms per cm. cube
Ohms per mil foot	0.06524	Microhms per in. cube
Poundals	0.03108	Pounds
Pounds	32.17	Poundals
Pound-feet	0.1383	Meter-Kilograms
Pounds of water	0.01602	Cubic feet
Pounds of water	0.1198	Gallons
Pounds per cubic foot	16.02	Kg. per cubic meter
Pounds per cubic foot	5.787×10^{-4}	Pounds per cubic in.
Pounds per cubic inch	27.68	Grams per cubic cm.
Pounds per cubic inch	2.768×10^{-4}	Kg. per cubic meter
Pounds per cubic inch	1.728	Pounds per cubic ft.
Pounds per square foot	0.01602	Feet of water
Pounds per square foot	4.882	Kg. per square meter

COMMONLY USED CONVERSION FACTORS (cont.)

Multiply	By	To Obtain
Pounds per square foot	6.944×10^{-3}	Pounds per sq. in.
Pounds per square inch	2.307	Feet of water
Pounds per square inch	2.036	Inches of mercury
Pounds per square inch	703.1	Kg. per square meter
Radians	57.30	Degrees
Square centimeters	1.973×10^{5}	Circular mils
Square Feet	2.296×10^{-5}	Acres
Square Feet	0.09290	Square meters
Square inches	1.273×10^{6}	Circular mils
Square inches	6.452	Square centimeters
Square Kilometers	0.3861	Square miles
Square meters	10.76	Square feet
Square miles	640	Acres
Square miles	2.590	Square kilometers
Square Millimeters	1.973×10^{3}	Circular mils
Square mils	1.273	Circular mils
Tons (long)	2,240	Pounds
Tons (metric)	2,205	Pounds
Tons (short)	2,000	Pounds
Watts	0.05686	B.t.u. per minute
Watts	10^{7}	Ergs per sec.
Watts	44.26	Foot-pounds per min.
Watts	1.341×10^{-3}	Horsepower
Watts	14.34	Calories per min.
Watts-hours	3.412	British thermal units
Watts-hours	2,655	Footpounds
Watts-hours	1.341×10^{-3}	Horsepower-hours
Watts-hours	0.8605	Kilogram-calories
Watts-hours	376.1	Kilogram-meters
Webers	10^{8}	Maxwells

ELECTRICAL PREFIXES

Prefixes

Prefixes are used to avoid long expressions of units that are smaller and larger than the base unit. See Common Prefixes. For example, sentences 1 and 2 do not use prefixes. Sentences 3 and 4 use prefixes.

1. A solid-state device draws 0.000001 amperes (A).
2. A generator produces 100,000 watts (W).
3. A solid-state device draws 1 microampere (uA).
4. A generator produces 100 kilowatts (kW).

Converting Units

To convert between different units, the decimal point is moved to the left or right, depending on the unit. See Conversion Table. For example, an electronic circuit has a current flow of .000001 A. The current value is converted to simplest terms by moving the decimal point six places to the right to obtain 1.0μA (from Conversion Table).

.000001. A = 1.0 uA

Move decimal point
6 places to right

Common Electrical Quantities

Abbreviations are used to simplify the expression of common electrical quantities.
See Common Electrical Quantities. For example, milliwatt is abbreviated mW, kilovolt is abbreviated kV, and ampere is abbreviated A.

COMMON PREFIXES

Symbol	Prefix	Equivalent
G	giga	1,000,000,000
M	mega	1,000,000
k	kilo	1000
base unit	—	1
m	milli	.001
u	micro	.000001
n	nano	.000000001

COMMON ELECTRICAL QUANTITIES

Variable	Name	Unit of Measure and Abbreviation
E	voltage	volt - V
I	current	ampere - A
R	resistance	ohm - Ω
P	power	watt - W
P	power (apparent)	volt-amp - VA
C	capacitance	farad - F
L	inductance	henry - H
Z	impedance	ohm - Ω
G	conductance	siemens - S
f	frequency	hertz - Hz
T	period	second - s

CONVERSION TABLE

Initial Units	Final Units						
	giga	mega	kilo	base unit	milli	micro	nano
giga	—	3R	6R	9R	12R	15R	18R
mega	3L	—	3R	6R	9R	12R	15R
kilo	6L	3L	—	3R	6R	9R	12R
base unit	9L	6L	3L	—	3R	6R	9R
milli	12L	9L	6L	3L	—	3R	6R
micro	15L	12L	9L	6L	3L	—	3R
nano	18L	15L	12L	9L	6L	3L	—

DECIMAL EQUIVALENTS OF FRACTIONS

8ths	**32nds**	**64ths**	**64ths**
$^{1}/_{8}$ = .125	$^{1}/_{32}$ = .03125	$^{1}/_{64}$ = 0.15625	$^{33}/_{64}$ = .515625
$^{1}/_{4}$ = .250	$^{3}/_{32}$ = .09375	$^{3}/_{64}$ = .046875	$^{35}/_{64}$ = .546875
$^{3}/_{8}$ = .375	$^{5}/_{32}$ = .15625	$^{5}/_{64}$ = .078125	$^{37}/_{64}$ = .578120
$^{1}/_{2}$ = .500	$^{7}/_{32}$ = .21875	$^{7}/_{64}$ = .109375	$^{39}/_{64}$ = .609375
$^{5}/_{8}$ = .625	$^{9}/_{32}$ = .28125	$^{9}/_{64}$ = .140625	$^{41}/_{64}$ = .640625
$^{3}/_{4}$ = .750	$^{11}/_{32}$ = .34375	$^{11}/_{64}$ = .171875	$^{43}/_{64}$ = .671875
$^{7}/_{8}$ = .875	$^{13}/_{32}$ = .40625	$^{13}/_{64}$ = .203128	$^{45}/_{64}$ = .703125
16ths	$^{15}/_{32}$ = .46875	$^{15}/_{64}$ = .234375	$^{47}/_{64}$ = .734375
$^{1}/_{16}$ = .0625	$^{17}/_{32}$ = .53125	$^{17}/_{64}$ = .265625	$^{49}/_{64}$ = .765625
$^{3}/_{16}$ = .1875	$^{19}/_{32}$ = .59375	$^{19}/_{64}$ = .296875	$^{51}/_{64}$ = 3796875
$^{5}/_{16}$ = .3125	$^{21}/_{32}$ = .65625	$^{21}/_{64}$ = .328125	$^{53}/_{64}$ = .828125
$^{7}/_{16}$ = .4375	$^{23}/_{32}$ = .71875	$^{23}/_{64}$ = .359375	$^{55}/_{64}$ = .859375
$^{9}/_{16}$ = .5625	$^{25}/_{32}$ = .78125	$^{25}/_{64}$ = .390625	$^{57}/_{64}$ = .890625
$^{11}/_{16}$ = .6875	$^{27}/_{32}$ = .84375	$^{27}/_{64}$ = .421875	$^{59}/_{64}$ = .921875
$^{13}/_{16}$ = .8125	$^{29}/_{32}$ = .90625	$^{29}/_{64}$ = .453125	$^{61}/_{64}$ = .953125
$^{15}/_{16}$ = .9375	$^{31}/_{32}$ = .96875	$^{31}/_{64}$ = .484375	$^{63}/_{64}$ = .984375

COMMON ENGINEERING UNITS AND THEIR RELATIONSHIP

Quantity	SI Metric Units/Symbols	Customary Units	Relationship of Units
Acceleration	meters per second squared (m/s^2)	feet per second squared (ft/s^2)	$m/s^2 = ft/s^2 \times 3.281$
Area	square meter (m^2) square millimeter (mm^2)	square foot (ft^2) square inch (in^2)	$m^2 = ft^2 \times 10.764$ $mm^2 = in^2 \times 0.00155$
Density	kilograms per cubic meter (kg/m^3) grams per cubic centimeter (g/cm^3)	pounds per cubic foot (lb/ft^3) pounds per cubic inch (lb/in^3)	$kg/m^3 = lb/ft^2 \times 16.02$ $g/cm^3 = lb/in^3 \times 0.036$
Work	Joule (J)	foot pound force (ft lbf or ft lb)	J = ft lbf x 1.356
Heat	Joule (J)	British thermal unit (Btu) Calorie (Cal)	J = Btu x 1.055 J = cal x 4.187
Energy	kilowatt (kW)	Horsepower (HP)	kW = HP x 0.7457
Force	Newton (N) Newton (N)	Pound-force (lbf, lb · f, or lb) kilogram-force (kgf, kg · f, or kp)	N = lbf x 4.448 $N = \frac{kgf}{9.807}$
Length	meter (m) millimeter (mm)	foot (ft) inch (in)	m = ft x 3.281 $mm = \frac{in}{25.4}$
Mass	kilogram (kg) gram (g)	pound (lb) ounce (oz)	kg = lb x 2.2 $g = \frac{oz}{28.35}$
Stress	Pascal = Newton per second (Pa = N/s)	pounds per square inch (lb/in^2 or psi)	$Pa = lb/in^2 \times 6{,}895$
Temperature	degree Celsius (°C)	degree Fahrenheit (°F)	$°C = \frac{°F - 32}{1.8}$
Torque	Newton meter (N · m)	foot-pound (ft lb) inch-pound (in lb)	N · m = ft lbf x 1.356 N · m = in lbf x 0.113
Volume	cubic meter (m^3) cubic centimeter (cm^3)	cubic foot (ft^3) cubic inch (in^3)	$m^3 = ft^3 \times 35.314$ $cm^3 = \frac{in^3}{16.387}$

TRIGONOMETRIC FORMULAS – RIGHT TRIANGLE

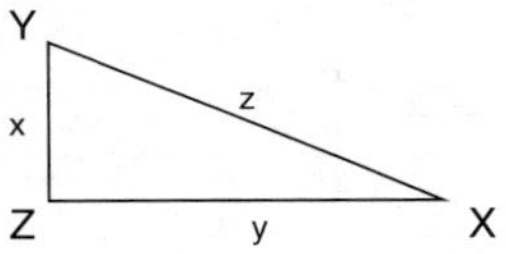

Angles = X, Y, Z
Distances = x, y, z
Area = $\frac{x\,y}{2}$

$\sin X = \frac{x}{z}$ $\cos X = \frac{y}{z}$

$\tan X = \frac{x}{y}$ $\cot X = \frac{y}{x}$

Pythagorean Theorem states
That $x^2 + y^2 = z^2$

Thus $x = \sqrt{z^2 - y^2}$

Thus $y = \sqrt{z^2 - x^2}$

Thus $z = \sqrt{x^2 + y^2}$

Given X and z, find Y, x and y

$Y = 90^\circ - X,\; x = z \sin X,\; y = z \cos X$

Given X and z, find Y, x and z

$Y = 90^\circ - X,\; x = y \tan X,\; z = \frac{y}{\cos X}$

Given x and z, find X, Y and y

$\sin X = \frac{x}{z} = \cos Y,\; y = \sqrt{(z^2 - x^2)} = z\sqrt{1 - \frac{x^2}{z^2}}$

Given x and y, find X, Y and z

$\tan X = \frac{x}{y} = \cot Y,\; z = \sqrt{x^2 + y^2} = x\sqrt{1 + \frac{y^2}{x^2}}$

Given X and x, find Y, y and z

$Y = 90^\circ - X,\; y = x \cot X,\; z = \frac{x}{\sin X}$

TRIGONOMETRIC FORMULAS – OBLIQUE TRIANGLES

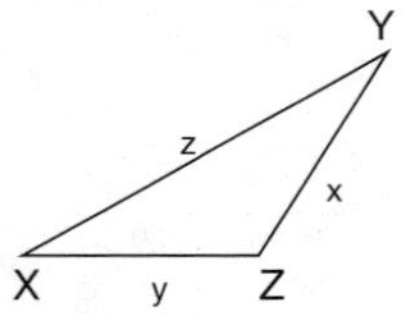

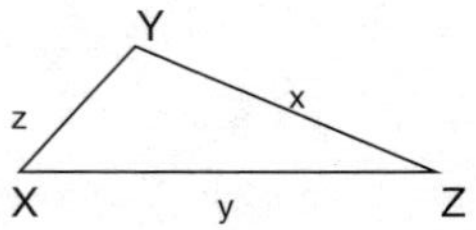

Given x, y and z, Find X, Y and Z

$$s = \frac{x+y+z}{2}, \quad \sin\frac{1}{2}X = \sqrt{\frac{(s-y)(s-z)}{yz}}$$

$$\sin\frac{1}{2}Y = \sqrt{\frac{(s-x)(s-z)}{xz}}, \quad C = 180^\circ - (X+Y)$$

Given x, y and z, find the Area

$$s = \frac{x+y+z}{2}, \quad \text{Area} = \sqrt{S(s-x)(s-y)(s-z)}$$

$$\text{Area} = \frac{yz \sin X}{2}, \quad \text{Area} = \frac{x^2 \sin Y \sin Z}{2 \sin X}$$

Given x, y, and Z, find X, Y and z

$$X + Y = 180^\circ - Z, \quad z = \frac{x \sin Z}{\sin X}, \quad \tan X = \frac{x \sin Z}{y - (x\cos Z)}$$

Given X, x and y, Find Y, Z and z

$$\sin Y = \frac{y \sin X}{x}, \quad Z = 180^\circ - (X+Y), \quad z = \frac{x \sin Z}{\sin X}$$

Given X, Y and x, Find y, Z and z

$$y = \frac{x \sin Y}{\sin X}, \quad Z = 180^\circ - (X+Y), \quad z = \frac{x \sin Z}{\sin X}$$

TRIGONOMETRIC FORMULAS – SHAPES

Equilateral Triangle

X = Sides (Equal Lengths)

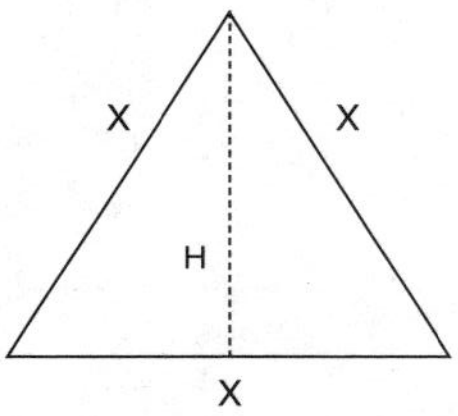

Area $= X^2 \sqrt{\frac{3}{4}} = .433\ X^2$

Perimeter $= 3\ X$

$H = \frac{X}{2} \sqrt{3} = .866\ X$

Annulus

C_1 and R_1 = Inside Circle

C_2 and R_2 = Outside Circle

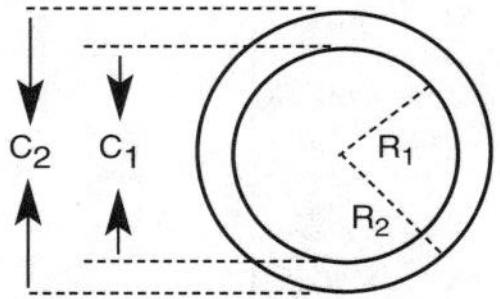

C = Circumference

R = Radius

Area $= \pi (R_1 + R_2)(R_2 - R_1)$

Area $= \left((C_2)^2 - (C_1)^2 \right) .7854$

Trapezium

Perimeter is the

Sum of L, M, N and O

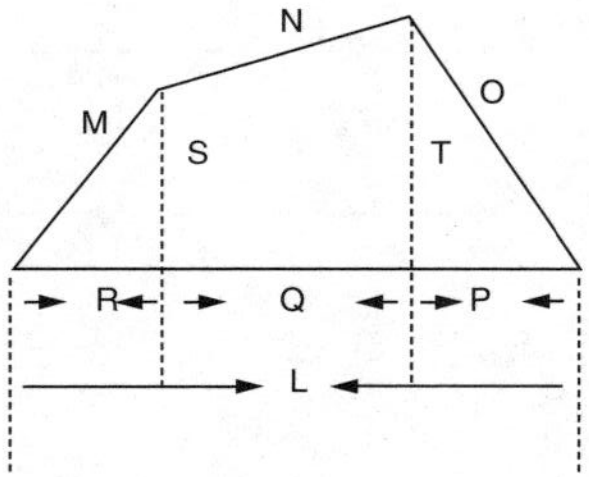

Area $= \frac{(S + T)\,Q + RS + PT}{2}$

TRIGONOMETRIC FORMULAS – SHAPES (cont.)

Trapezoid

Perimeter =
The Sum of the lengths of all four sides

Area = $\frac{(X + Y)}{2}$

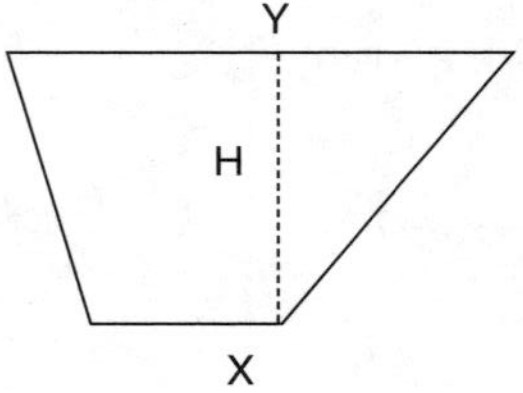

Quadrilateral

Area = $\frac{L_1 \bullet L_2 \bullet \sin\theta}{2}$

Where θ =
Degrees
of Angle

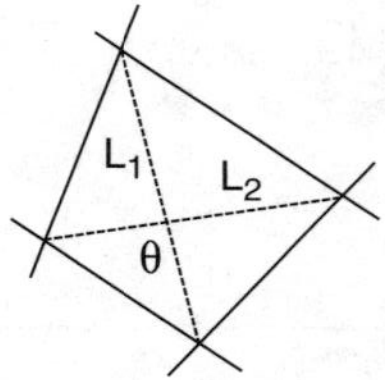

Rectangle

Area = XY
Diagonal Line (D)

= $\sqrt{X^2 + Y^2}$

Perimeter =
$2(X + Y)$
If a square
then X = Y

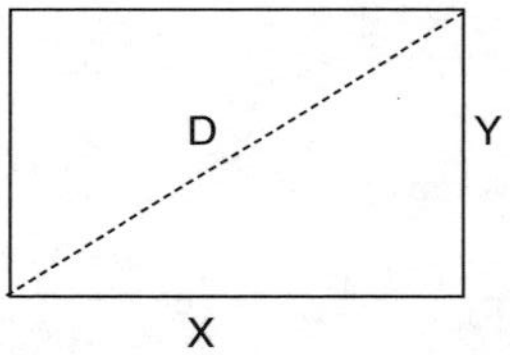

Parallelogram

Where θ =
Degrees
of Angle

Area =
XH = XY sin θ
Perimeter =
$2(X + Y)$

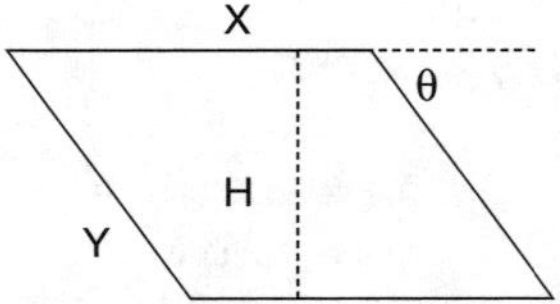

SHEET METAL SCREW CHARACTERISTICS				
Screw Size #	Screw Dia. (Inches)	Diameter of Pierced Hole (Inches)	Hole Size	Thickness of Metal – Gauge #
#4	.112	.086	#44	28
		.086	#44	26
		.093	#42	24
		.098	#42	22
		.100	#40	20
#6	.138	.111	#39	28
		.111	#39	26
		.111	#39	24
		.111	#38	22
		.111	#36	20
#7	.155	.121	#37	28
		.121	#37	26
		.121	#35	24
		.121	#33	22
		.121	#32	20
		–	#31	18
#8	.165	.137	#33	26
		.137	#33	24
		.137	#32	22
		.137	#31	20
		–	#30	18
#10	.191	.158	#30	26
		.158	#30	24
		.158	#30	22
		.158	#29	20
		.158	#25	18
#12	.218	–	#26	24
		.185	#25	22
		.185	#24	20
		.185	#22	18
#14	.251	–	#15	24
		.212	#12	22
		.212	#11	20
		.212	#9	18
Deviations in materials and conditions could require variations from these dimensions				

ALLEN HEAD AND MACHINE SCREW BOLT & TORQUE CHARACTERISTICS

Number of Threads Per Inch	Allen Head And Mach. Screw Bolt Size	Allen Head Case H Steel 160,000 psi	Mach. Screw Yellow Brass 60,000 psi	Mach. Screw Silicone Bronze 70,000 psi
		Torque in Foot-Pounds or Inch-Pounds		
4.5	2"	8800	–	–
5	1 3/4"	6100	–	–
6	1 1/2"	3450	655	595
6	1 3/8"	2850	–	–
7	1 1/4"	2130	450	400
7	1 1/8"	1520	365	325
8	1"	970	250	215
9	7/8"	640	180	160
10	3/4"	400	117	104
11	5/8"	250	88	78
12	9/16"	180	53	49
13	1/2"	125	41	37
14	7/16"	84	30	27
16	3/8"	54	20	17
18	5/16"	33	125 in#	110 in#
20	1/4"	16	70 in#	65 in#
24	#10	60	22 in#	20 in#
32	#8	46	19 in#	16 in#
32	#6	21	10 in#	8 in#
40	#5	–	7.2 in#	6.4 in#
40	#4	–	4.9 in#	4.4 in#
48	#3	–	3.7 in#	3.3 in#
56	#2	–	2.3 in#	2 in#

For fine thread bolts, increase by 9%.

HEX HEAD BOLT & TORQUE CHARACTERISTICS

BOLT MAKE-UP IS STEEL WITH COARSE THREADS

Number of Threads Per Inch	Hex Head Bolt Size	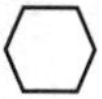 SAE 0-1-2 74,000 psi	 SAE Grade 3 100,000 psi	 SAE Grade 5 120,000 psi
		Torque = Foot-Pounds		
4.5	2"	2750	5427	4550
5	1 3/4"	1900	3436	3150
6	1 1/2"	1100	1943	1775
6	1 3/8"	900	1624	1500
7	1 1/4"	675	1211	1105
7	1 1/8"	480	872	794
8	1"	310	551	587
9	7/8"	206	372	382
10	3/4"	155	234	257
11	5/8"	96	145	154
12	9/16"	69	103	114
13	1/2"	47	69	78
14	7/16"	32	47	54
16	3/8"	20	30	33
18	5/16"	12	17	19
20	1/4"	6	9	10

For fine thread bolts, increase by 9%.

HEX HEAD BOLT & TORQUE CHARACTERISTICS (cont.)

BOLT MAKE-UP IS STEEL WITH COARSE THREADS

Number of Threads Per Inch	Hex Head Bolt Size	SAE Grade 6 133,000 psi	SAE Grade 7 133,000 psi	SAE Grade 8 150,000 psi
		Torque = Foot-Pounds		
4.5	2"	7491	7500	8200
5	1 3/4"	5189	5300	5650
6	1 1/2"	2913	3000	3200
6	1 3/8"	2434	2500	2650
7	1 1/4"	1815	1825	1975
7	1 1/8"	1304	1325	1430
8	1"	825	840	700
9	7/8"	550	570	600
10	3/4"	350	360	380
11	5/8"	209	215	230
12	9/16"	150	154	169
13	1/2"	106	110	119
14	7/16"	69	71	78
16	3/8"	43	44	47
18	5/16"	24	25	29
20	1/4"	12.5	13	14

For fine thread bolts, increase by 9%.
For special alloy bolts, obtain torque rating from the manufacturer.

WHITWORTH HEX HEAD BOLT & TORQUE CHARACTERISTICS

BOLT MAKE-UP IS STEEL WITH COARSE THREADS

Number of Threads Per Inch	Whitworth Type Hex Head Bolt Size	Grades A & B 62,720 psi	Grade S 112,000 psi	Grade T 123,200 psi	Grade V 145,600 psi
		Torque = Foot-Pounds			
8	1"	276	497	611	693
9	7/8"	186	322	407	459
11	3/4"	118	213	259	287
11	5/8"	73	128	155	175
12	9/16"	52	94	111	128
12	1/2"	36	64	79	89
14	7/16"	24	43	51	58
16	3/8"	15	27	31	36
18	5/16"	9	15	18	21
20	1/4"	5	7	9	10

For fine thread bolts, increase by 9%.

METRIC HEX HEAD BOLT & TORQUE CHARACTERISTICS

BOLT MAKE-UP IS STEEL WITH COARSE THREADS

(Metric Type) Thread Pitch	(Dimensions in Millimeters) Bolt Size	5D Standard 5D 71,160 psi	8G Standard 8G 113,800 psi	10K Standard 10K 142,000 psi	12K Standard 12K 170,674 psi
		Torque = Foot-Pounds			
3.0	24	261	419	570	689
2.5	22	182	284	394	464
2.0	18	111	182	236	183
2.0	16	83	132	175	208
1.25	14	55	89	117	137
1.25	12	34	54	70	86
1.25	10	19	31	40	49
1.0	8	10	16	22	27
1.0	6	5	6	8	10

For fine thread bolts, increase by 9%.

TIGHTENING TORQUE IN POUND-FEET-SCREW FIT

Wire Size, AWG/kcmil	Driver	Bolt	Other
18-16	1.67	6.25	4.2
14-8	1.67	6.25	6.125
6-4	3.0	12.5	8.0
3-1	3.2	21.00	10.40
0-2/0	4.22	29	12.5
3/0-200	–	37.5	17.0
250-300	–	50.0	21.0
400	–	62.5	21.0
500	–	62.5	25.0
600-750	–	75.0	25.0
800-1000	–	83.25	33.0
1250-2000	–	83.26	42.0

SCREW TORQUES

Screw Size, Inches Across, Hex Flats	Torque, Pound-Feet
1/8	4.2
5/32	8.3
3/16	15
7/32	23.25
1/4	42

STANDARD TAPS AND DIES (IN INCHES)

Thread Size	Coarse			Fine		
	Drill Size	Threads Per In.	Decimal Size	Drill Size	Threads Per In.	Decimal Size
4"	3	4	3.75			
3 3/4"	3	4	3.5			
3 1/2"	3	4	3.25			
3 1/4"	3	4	3.0			
3"	2	4	2.75			
2 3/4"	2	4	2.5			
2 1/2"	2	4	2.25	–	–	–
2 1/4"	2	4.5	2.0313	–	–	–
2"	1	4.5	1.7813	–	–	–
1 3/4"	1	2	1.5469	–	–	–
1 1/2"	1	6	1.3281	1 27/64"	12	1.4219
1 3/8"	1	6	1.2188	1 19/64"	12	1.2969
1 1/4"	1	7	1.1094	1 11/64"	12	1.1719
1 1/8"	63/64"	7	.9844	1 3/64"	12	1.0469
1"	7/8"	8	.8750	15/16"	14	.9375
7/8"	49/64"	9	.7656	13/16"	14	.8125
3/4"	21/32"	10	.6563	11/16"	16	.6875
5/8"	17/32"	11	.5313	37/64"	18	.5781
9/16"	31/64"	12	.4844	33/64"	18	.5156
1/2"	27/64"	13	.4219	29/64"	20	.4531
7/16"	U	14	.368	25/64"	20	.3906
3/8"	5/16"	16	.3125	Q	24	.332
5/16"	F	18	.2570	I	24	.272
1/4"	#7	20	.201	#3	28	.213
#12	#16	24	.177	#14	28	.182
#10	#25	24	.1495	#21	32	.159
3/16"	#26	24	.147	#22	32	.157
#8	#29	32	.136	#29	36	.136
#6	#36	32	.1065	#33	40	.113
#5	#38	40	.1015	#37	44	.104
1/8"	3/32"	32	.0938	#38	40	.1015
#4	#43	40	.089	#42	48	.0935
#3	#47	48	.0785	#45	56	.082
#2	#50	56	.07	#50	64	.07
#1	#53	64	.0595	#53	72	.0595
#0	–	–	–	3/64"	80	.0469

(mm) Thread Pitch	TAPS & DIES – METRIC CONVERSIONS			
	Fine Thread Size		Tap Drill Size	
	Inches	mm	Inches	mm
4.5	1.6535	42	1.4567	37.0
4.0	1.5748	40	1.4173	36.0
4.0	1.5354	39	1.3779	35.0
4.0	1.4961	38	1.3386	34.0
4.0	1.4173	36	1.2598	32.0
3.5	1.3386	34	1.2008	30.5
3.5	1.2992	33	1.1614	29.5
3.5	1.2598	32	1.1220	28.5
3.5	1.1811	30	1.0433	26.5
3.0	1.1024	28	.9842	25.0
3.0	1.0630	27	.9449	24.0
3.0	1.0236	26	.9055	23.0
3.0	.9449	24	.8268	21.0
2.5	.8771	22	.7677	19.5
2.5	.7974	20	.6890	17.5
2.5	.7087	18	.6102	15.5
2.0	.6299	16	.5118	14.0
2.0	.5512	14	.4724	12.0
1.75	.4624	12	.4134	10.5
1.50	.4624	12	.4134	10.5
1.50	.3937	11	.3780	9.6
1.50	.3937	10	.3386	8.6
1.25	.3543	9	.3071	7.8
1.25	.3150	8	.2677	6.8
1.0	.2856	7	.2362	6.0
1.0	.2362	6	.1968	5.0
.90	.2165	5.5	.1811	4.6
.80	.1968	5	.1653	4.2
.75	.1772	4.5	.1476	3.75
.70	.1575	4	.1299	3.3
.75	.1575	4	.1279	3.25
.60	.1378	3.5	.1142	2.9
.60	.1181	3	.0945	2.4
.50	.1181	3	.0984	2.5
.45	.1124	2.6	.0827	2.1
.45	.0984	2.5	.0787	2.0
.40	.0895	2.3	.0748	1.9
.40	.0787	2	.0630	1.6
.45	.0787	2	.0590	1.5
.35	.0590	1.5	.0433	1.1

RECOMMENDED DRILLING SPEEDS (RPMS)

Material	Bit Sizes	RPM Speed Range
Glass	Special Metal Tube Drilling	700
Plastics	7/16" and larger	500 – 1000
	3/8"	1500 – 2000
	5/16"	2000 – 2500
	1/4"	3000 – 3500
	3/16"	3500 – 4000
	1/8"	5000 – 6000
	1/16" and smaller	6000 – 6500
Woods	1" and larger	700 – 2000
	3/4" to 1"	2000 – 2300
	1/2" to 3/4"	2300 – 3100
	1/4" to 1/2"	3100 – 3800
	1/4" and smaller	3800 – 4000
	carving / routing	4000 – 6000
Soft Metals	7/16" and larger	1500 – 2500
	3/8"	3000 – 3500
	5/16"	3500 – 4000
	1/4"	4500 – 5000
	3/16"	5000 – 6000
	1/8"	6000 – 6500
	1/16" and smaller	6000 – 6500
Steel	7/16" and larger	500 – 1000
	3/8"	1000 – 1500
	5/16"	1000 – 1500
	1/4"	1500 – 2000
	3/16"	2000 – 2500
	1/8"	3000 – 4000
	1/16" and smaller	5000 – 6500
Cast Iron	7/16" and larger	1000 – 1500
	3/8"	1500 – 2000
	5/16"	1500 – 2000
	1/4"	2000 – 2500
	3/16"	2500 – 3000
	1/8"	3500 – 4500
	1/16" and smaller	6000 – 6500

TORQUE LUBRICATION EFFECTS IN FOOT-POUNDS

Lubricant	5/16" – 18 Thread	1/2" – 13 Thread	Torque Decrease
Graphite	13	62	49 - 55%
Mily Film	14	66	45 - 52%
White Grease	16	79	35 - 45%
Sae 30	16	79	35 - 45%
Sae 40	17	83	31 - 41%
Sae 20	18	87	28 - 38%
Plated	19	90	26 - 34%
No Lube	29	121	0%

METALWORKING LUBRICANTS

Materials	Threading	Lathing	Drilling
Machine Steels	Dissolvable Oil Mineral Oil Lard Oil	Dissolvable Oil	Dissolvable Oil Sulpherized Oil Min. Lard Oil
Tool Steels	Lard Oil Sulpherized Oil	Dissolvable Oil	Dissolvable Oil Sulpherized Oil
Cast Irons	Sulpherized Oil Dry Min. Lard Oil	Dissolvable Oil Dry	Dissolvable Oil Dry Air Jet
Malleable Irons	Soda Water Lard Oil	Soda Water Dissolvable Oil	Soda Water Dry
Aluminums	Kerosene Dissolvable Oil Lard Oil	Dissolvable Oil	Kerosene Dissolvable Oil
Brasses	Dissolvable Oil Lard Oil	Dissolvable Oil	Kerosene Dissolvable Oil Dry
Bronzes	Dissolvable Oil Lard Oil	Dissolvable Oil	Dissolvable Oil Dry
Coppers	Dissolvable Oil Lard Oil	Dissolvable Oil	Kerosene Dissolvable Oil Dry

TYPES OF SOLDERING FLUX

To Solder	Use
For cast iron	Cuprous oxide
For galvanized iron, galvanized, steel, tin, zinc	Hydrochloric acid
For pewter and lead	Organic
For brass, copper, gold, iron, silver, steel	Borax
For brass, bronze, cadmium, copper, lead, silver	Resin
For brass, copper, gun metal, iron, nickel, tin, zinc	Ammonia chloride
For bismuth, brass, copper, gold, silver, tin	Zinc chloride
For silver	Sterling
For pewter and lead	Tallow
For stainless only	Stainless steel (only)

HARD SOLDER ALLOYS

To hard solder	Copper %	Gold %	Silver %	Zinc %
Gold	22	67	11	
Silver	20		70	10
Hard brass	45			55
Soft brass	22			78
Copper	50			50
Cast iron	55			45
Steel and iron	64			36

SOFT SOLDER ALLOYS

To soft solder	Lead %	Tin %	Zinc %	Bism %	Other %
Gold	33	67			
Silver	33	67			
Brass	34	66			
Copper	40	60			
Steel and iron	50	50			
Galvanized steel	42	58			
Tinned steel	36	64			
Zinc	45	55			
Block Tin	1	99			
Lead	67	33			
Gun metal	37	63			
Pewter	25	25		50	
Bismuth	33	33		34	
Aluminum		70	25		5

PROPERTIES OF WELDING GASES

Type of Gas	Characteristics	Common Tank Sizes (cu. ft.)
Acetylene	C_2H_2, explosive gas, flammable, garlic-like odor, colorless, dangerous if used in pressures over 15 psig (30 psig absolute)	10, 40, 75 100, 300
Argon	Ar, non-explosive inert gas, tasteless, odorless, colorless	131, 330 4754 (liquid)
Carbon Dioxide	CO_2, Non-explosive inert gas, tasteless, odorless, colorless (in large quantities is toxic)	20 lbs., 50 lbs.
Helium	He, Non-explosive inert gas, tasteless, odorless, colorless	221
Hydrogen	H_2, explosive gas, tasteless, odorless, colorless	191
Nitrogen	N_2, Non-explosive inert gas, tasteless, odorless, colorless	20, 40, 80 113, 225
Oxygen	O_2, Non-explosive gas, tasteless, odorless, colorless, supports combustion	20, 40, 80 122, 244 4500 (liquid)

WELDING RODS – 36" LONG

Rod Size (Dia)	Number of Rods Per Pound			
	Aluminum	Brass	Cast Iron	Steel
3/8"	–	1.0	.25	1.0
5/16"	–	–	.50	1.33
1/4"	6.0	2.0	2.25	2.0
3/16"	9.0	3.0	5.50	3.5
5/32"	–	–	–	5.0
1/8"	23.0	7.0	–	8.0
3/32"	41.0	13.0	–	14.0
1/16"	91.0	29.0	–	31.0

STRENGTH GAIN VS. PULL ANGLE

The weight bearing capacity of a strap, for example, increases by the factor K shown below as the angle of the strap decreases. A 100 lb. capacity strap at a 60° pulling angle loses 50 lbs. of weight bearing ability.
At perfectly vertical, the ability is 100%.

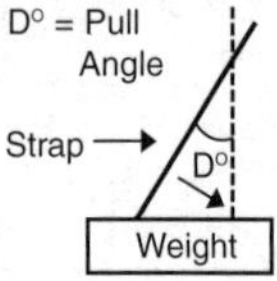

K	D°	K	D°	K	D°
.7412	75	.3572	50	.0937	25
.6580	70	.2929	45	.0603	20
.5774	65	.2340	40	.0341	15
.5000	60	.1340	35	.0152	10
.4264	55	.1208	30	.0038	5

LENGTH OF WIRE CABLE PER REEL

Cable length (feet) = X (X + Y) Z (K)
Use equation above to determine the length of wire cable that is wound smoothly on a reel. The dimensions for X, Y and Z are in inches.

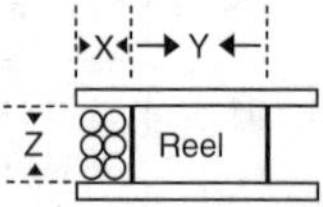

(K)	Cable Dia. (Inches)	(K)	Cable Dia. (Inches)
.0476	2.250	.239	1.000
.0532	2.125	.308	.875
.0597	2.000	.428	.750
.0675	1.875	.607	.625
.0770	1.750	.741	.563
.0886	1.625	.925	.500
.107	1.500	1.19	.438
.127	1.375	1.58	.375
.152	1.250	2.21	.313
.191	1.125	3.29	.250

STEEL WIRE DIAMETERS

Gauge #	American or Brown & Sharp	US Steelwire or Washburn & Moen	Birmingham or Stubs Iron	British Imperial Std Wire
7/0	–	0.4900	–	0.500
6/0	0.580000	0.4615	–	0.464
5/0	0.516500	0.4305	–	0.432
4/0	0.460000	0.3938	0.454	0.400
3/0	0.409642	0.3625	0.425	0.372
2/0	0.364796	0.3310	0.380	0.348
0	0.324861	0.3065	0.340	0.324
1	0.289297	0.2830	0.300	0.300
2	0.257627	0.2625	0.284	0.276
3	0.229423	0.2437	0.259	0.252
4	0.204307	0.2253	0.238	0.232
5	0.181940	0.2070	0.220	0.212
6	0.162023	0.1920	0.203	0.192
7	0.144285	0.1770	0.180	0.176
8	0.128490	0.1620	0.165	0.160
9	0.114423	0.1483	0.148	0.144
10	0.101897	0.1350	0.134	0.128
11	0.090742	0.1205	0.120	0.116
12	0.080808	0.1055	0.109	0.104
13	0.071962	0.0915	0.095	0.092
14	0.064084	0.0800	0.083	0.080
15	0.057068	0.0720	0.072	0.072
16	0.050821	0.0625	0.065	0.064
17	0.045257	0.0540	0.058	0.056

STEEL WIRE DIAMETERS (cont.)

Gauge #	American or Brown & Sharp	US Steelwire or Washburn & Moen	Birmingham or Stubs Iron	British Imperial Std Wire
18	0.040303	0.0475	0.049	0.048
19	0.035890	0.0410	0.042	0.040
20	0.031961	0.0348	0.035	0.036
21	0.028462	0.03175	0.032	0.032
22	0.025346	0.0286	0.028	0.028
23	0.022572	0.0258	0.025	0.024
24	0.020101	0.0230	0.022	0.022
25	0.017900	0.0204	0.020	0.020
26	0.015941	0.0181	0.018	0.018
27	0.014195	0.0173	0.016	0.0164
28	0.012641	0.0162	0.014	0.0148
29	0.011257	0.0150	0.013	0.0136
30	0.010025	0.0140	0.012	0.0124
31	0.008928	0.0132	0.010	0.0116
32	0.007950	0.0128	0.009	0.0108
33	0.007080	0.0118	0.008	0.0100
34	0.006305	0.0104	0.007	0.0092
35	0.005615	0.0095	0.005	0.0084
36	0.005000	0.0090	0.004	0.0076
37	0.004453	0.0085	–	0.0068
38	0.003965	0.0080	–	0.0060
39	0.003531	0.0075	–	0.0052
40	0.003144	0.0070	–	0.0048

WIRE ROPE CHARACTERISTICS FOR 6 STRAND BY 19 WIRE TYPE			
Diameter in inches	**Weight in lbs./foot**	**Breaking Point in lbs.**	**Safe Load in lbs.**
1/4	0.10	4800	675
5/16	0.16	7400	1000
3/8	0.23	10600	1500
7/16	0.31	14400	2000
1/2	0.40	18700	2400
9/16	0.51	23600	3300
5/8	0.63	29000	4000
3/4	0.90	41400	6000
7/8	1.23	56000	8000
1	1.60	72800	10000
1 1/8	2.03	91400	13000
1 1/4	2.50	112400	16000
1 3/8	3.03	135000	19000
1 1/2	3.60	160000	22000
1 3/4	4.90	216000	30500
2	6.40	278000	40000
2 1/2	10.00	424000	60000

The above values are for vertical pulls at average ambient temperatures.

CABLE CLAMPS PER WIRE ROPE SIZE

Wire Rope Diameter (inches)	# of Clamps Required	Clip Spacing (inches)	Rope Turn-back (inches)
1/8	2	3	3 1/4
3/16	2	3	3 3/4
1/4	2	3 1/4	4 3/4
5/16	2	3 1/4	5 1/4
3/8	2	4	6 1/2
7/16	2	4 1/2	4
1/2	3	5	11 1/2
9/16	3	5 1/2	12
5/8	3	5 3/4	12
3/4	4	6 3/4	18
7/8	4	8	19
1	5	8 3/4	26
1 1/8	6	9 3/4	34
1 1/4	6	10 3/4	37
1 7/16	7	11 1/2	44
1 1/2	7	12 1/2	48
1 5/8	7	13 1/4	51
1 3/4	7	14 1/2	53
2	8	16 1/2	71
2 1/4	8	16 1/2	73
2 1/2	9	17 3/4	84
2 3/4	10	18	100
3	10	18	106

ROPE CHARACTERISTICS

Diameter in inches	Safe Load Ratio	Nylon Break Lbs.	Nylon Lbs./100 Feet	Polypropylene Break Lbs.	Polypropylene Lbs./100 Feet	Manila Break Lbs.	Manila Lbs./100 Feet
3/16	10:1	1000	1.0	800	0.7	406	1.5
1/4	10:1	1650	1.5	1250	1.2	540	2.0
5/16	10:1	2550	2.5	1900	1.8	900	2.9
3/8	10:1	3700	3.5	2700	2.8	1220	4.1
7/16	10:1	5000	5.0	3500	3.8	1580	5.3
1/2	9:1	6400	6.5	4200	4.7	2380	7.5
9/16	8:1	8000	8.3	5100	6.1	3100	10.4
5/8	8:1	10400	10.5	6200	7.5	3960	13.3
3/4	7:1	14200	14.5	8500	10.7	4860	16.7
13/16	7:1	17000	17.0	9900	12.7	5850	19.5
7/8	7:1	20000	20.0	11500	15.0	6950	22.4
1	7:1	25000	26.4	14000	18.0	8100	27.0
1 1/16	7:1	28800	29.0	16000	20.4	9450	31.2
1 1/8	7:1	33000	34.0	18300	23.8	10800	36.0
1 1/4	7:1	37500	40.0	21000	27.0	12200	41.6
1 5/16	7:1	43000	45.0	23500	30.4	13500	47.8
1 1/2	7:1	53000	55.0	29700	38.4	16700	60.0
1 5/8	7:1	65000	66.5	36000	47.6	20200	74.5
1 3/4	7:1	78000	83.0	43000	59.0	23800	89.5
2	7:1	92000	95.0	52000	69.0	28000	108
2 1/8	7:1	106000	109	61000	80.0		
2 1/4	6:1	125000	129	69000	92.0		
2 1/2	6:1	140000	149	80000	107		
2 5/8	6:1	162000	168	90000	120		
2 7/8	6:1	180000	189	101000	137		
3	6:1	200000	210	114000	153		
3 1/4	6:1	250000	264	137000	190		
3 1/2	6:1	300000	312	162000	232		
4	6:1	360000	380	190000	276		

Lbs/foot = Rope weight per linear foot
Break Lbs = Tensile strength
Safe Load Ratio = Break strength to safe load
Example: 7/16" nylon rope break strength = 5000 lbs.
5000/10 = 500 lbs. safe working load

Note: increased temperatures decrease rope strength

CHAIN CHARACTERISTICS

Rod Diameter in inches	Weight in lbs./foot	Breaking Point in lbs.	Safe Load in lbs.
1/4	0.75	5000	1200
5/16	1	7000	1700
3/8	1.5	10000	2500
7/16	2	14000	3500
1/2	2.5	18000	4500
9/16	3.25	22000	5500
5/8	4	27000	6700
11/16	5	32500	8100
3/4	6.25	40000	10000
13/16	7	42000	10500
7/8	8	48000	12000
15/16	9	54000	13500
1	10	61000	15200
1 1/16	12	69000	17200
1 1/8	13	78000	19500
1 3/16	14.5	88000	22000
1 1/4	16	95000	23700
1 5/16	17.5	104000	26000
1 3/8	19	114000	28500
1 7/16	21.5	122000	30500
1 1/2	23	134000	33500
1 9/16	25	142000	35500
1 5/8	28	154000	38500
1 11/16	30	158000	39500
1 3/4	31	166000	41500
1 13/16	33	178000	44500
1 7/8	35	190000	47500
1 15/16	38	202000	50500
2	40	216000	54000
2 1/4	53	273000	68200
2 1/2	65	337000	84200
2 3/4	73	387000	96700
3	86	436000	109000

The above values are based on vertical pulls at average ambient temperatures.

STEEL SHEET MEASUREMENTS

Gauge #	Lbs. per sq. ft.	Thickness (inches)		Weight lbs./sq. ft.	
		US Standard Gauge	Manufacturers Standard	Galvanized Sheet	Stainless Steel
7/0	20.00	0.5000			
6/0	18.75	0.4687			
5/0	17.50	0.4375			
4/0	16.25	0.4062			
3/0	15.00	0.3750			
2/0	13.75	0.3437	–		
0	12.50	0.3125	–		
1	11.25	0.2812	–		
2	10.62	0.2656	–		
3	10.00	0.2500	0.2391		
4	9.37	0.2344	0.2242		
5	8.75	0.2187	0.2092		
6	8.12	0.2031	0.1943		
7	7.50	0.1875	0.1793		
8	6.87	0.1719	0.1644		
9	6.25	0.1562	0.1495	–	–
10	5.62	0.1406	0.1345	5.7812	5.7937
11	5.00	0.1250	0.1196	5.1562	5.1500
12	4.37	0.1094	0.1046	4.5312	4.5063
13	3.75	0.0937	0.0897	3.9062	3.8625
14	3.12	0.0781	0.0747	3.2812	3.2187
15	2.81	0.0703	0.0673	2.9687	2.8968
16	2.50	0.0625	0.0598	2.6562	2.5750
17	2.25	0.0562	0.0538	2.4062	2.3175
18	2.00	0.0500	0.0478	2.1562	2.0600
19	1.75	0.0437	0.0418	1.9062	1.8025

STEEL SHEET MEASUREMENTS (cont.)

Gauge #	Lbs. per sq. ft.	Thickness (inches)		Weight lbs./sq. ft.	
		US Standard Gauge	Manufacturers Standard	Galvanized Sheet	Stainless Steel
20	1.50	0.0375	0.0359	1.6562	1.5450
21	1.37	0.0344	0.0329	1.5312	1.4160
22	1.25	0.0312	0.0299	1.4062	1.2875
23	1.12	0.0281	0.0269	1.2812	1.1587
24	1.00	0.0250	0.0239	1.1562	1.0300
25	0.875	0.0219	0.0209	1.0312	0.9013
26	0.750	0.0187	0.0179	0.9062	0.7725
27	0.687	0.0172	0.0164	0.8437	0.7081
28	0.625	0.0156	0.0149	0.7812	0.6438
29	0.562	0.0141	0.0135	0.7187	0.5794
30	0.500	0.0125	0.0120	0.6562	0.5150
31	0.437	0.0109	0.0105	–	–
32	0.406	0.0102	0.0097	–	–
33	0.375	0.0094	0.0090	–	–
34	0.344	0.0086	0.0082	–	–
35	0.312	0.0078	0.0075		
36	0.281	0.0070	0.0067		
37	0.266	0.0066	0.0064		
38	0.250	0.0062	0.0060		
39	0.234	0.0059	–		
40	0.219	0.0055			
41	0.211	0.0053			
42	0.203	0.0051			
43	0.195	0.0049			
44	0.187	0.0047			

STEEL PLATE WEIGHTS & SIZES

Thickness in inches	Weight in pounds per square foot
3/16	7.65
1/4	10.20
5/16	12.75
3/8	15.30
7/16	17.85
1/2	20.40
9/16	22.95
5/8	25.50
11/16	28.05
3/4	30.60
13/16	33.15
7/8	35.70
1	40.80
1 1/8	45.90
1 1/4	51.00
1 3/8	56.10
1 1/2	61.20
1 5/8	66.30
1 3/4	71.40
1 7/8	76.50
2	81.60
2 1/8	86.70
2 1/4	91.80
2 1/2	102.00
2 3/4	112.20
3	122.40
3 1/4	132.60
3 1/2	142.80
3 3/4	153.00
4	163.20
4 1/4	173.40
4 1/2	183.60
5	204.00
5 1/2	224.40
6	244.80
6 1/2	265.20
7	285.60
7 1/2	306.00
8	326.40
9	367.20
10	408.00

About The Author

Paul Rosenberg has an extensive background in the construction, data, electrical, HVAC and plumbing trades. He is a leading voice in the electrical industry with years of experience from an apprentice to a project manager. Paul has written for all of the leading electrical and low voltage industry magazines and has authored more than 30 books.

In addition, he wrote the first standard for the installation of optical cables (ANSI-NEIS-301) and was awarded a patent for a power transmission module. Paul currently serves as contributing editor for *Power Outlet Magazine*, teaches for Iowa State University and works as a consultant and expert witness in legal cases. He speaks occasionally at industry events.